Eszter Kapusi

Elimination of a selectable marker gene in barley

Eszter Kapusi

Elimination of a selectable marker gene in barley

via segregation of uncoupled T-DNAs in populations of doubled haploids

Südwestdeutscher Verlag für Hochschulschriften

Impressum / Imprint

Bibliografische Information der Deutschen Nationalbibliothek: Die Deutsche Nationalbibliothek verzeichnet diese Publikation in der Deutschen Nationalbibliografie; detaillierte bibliografische Daten sind im Internet über http://dnb.d-nb.de abrufbar.

Alle in diesem Buch genannten Marken und Produktnamen unterliegen warenzeichen-, marken- oder patentrechtlichem Schutz bzw. sind Warenzeichen oder eingetragene Warenzeichen der jeweiligen Inhaber. Die Wiedergabe von Marken, Produktnamen, Gebrauchsnamen, Handelsnamen, Warenbezeichnungen u.s.w. in diesem Werk berechtigt auch ohne besondere Kennzeichnung nicht zu der Annahme, dass solche Namen im Sinne der Warenzeichen- und Markenschutzgesetzgebung als frei zu betrachten wären und daher von jedermann benutzt werden dürften.

Bibliographic information published by the Deutsche Nationalbibliothek: The Deutsche Nationalbibliothek lists this publication in the Deutsche Nationalbibliografie; detailed bibliographic data are available in the Internet at http://dnb.d-nb.de.

Any brand names and product names mentioned in this book are subject to trademark, brand or patent protection and are trademarks or registered trademarks of their respective holders. The use of brand names, product names, common names, trade names, product descriptions etc. even without a particular marking in this works is in no way to be construed to mean that such names may be regarded as unrestricted in respect of trademark and brand protection legislation and could thus be used by anyone.

Coverbild / Cover image: www.ingimage.com

Verlag / Publisher:
Südwestdeutscher Verlag für Hochschulschriften
ist ein Imprint der / is a trademark of
AV Akademikerverlag GmbH & Co. KG
Heinrich-Böcking-Str. 6-8, 66121 Saarbrücken, Deutschland / Germany
Email: info@svh-verlag.de

Herstellung: siehe letzte Seite /
Printed at: see last page
ISBN: 978-3-8381-3565-6

Zugl. / Approved by: Halle (Saale), Martin Luther University of Halle-Wittenberg, Diss., 2010

Copyright © 2012 AV Akademikerverlag GmbH & Co. KG
Alle Rechte vorbehalten. / All rights reserved. Saarbrücken 2012

Table of contents

1. INTRODUCTION ... 7
1.1. Agroeconomic importance of barley ... 7
1.2. Genetic transformation of barley ... 9
   1.2.1 Methods of gene transfer ... 9
   1.2.2 Vector systems ... 10
   1.2.3. Integration of T-DNA sequences ... 11
   1.2.4. Selectable markers and reporter genes ... 13
   1.2.5. Generation of marker-free lines ... 15
1.3. Haploid technology ... 18
   1.3.1. Generation of doubled haploid barley ... 19
   1.3.2. Simplified segregation of transgenes in populations of doubled haploids ... 20
1.4. Scope of the thesis ... 21

2. MATERIALS AND METHODS ... 23
2.1. Bacterial strains and vectors ... 23
   2.1.1. Bacterial media and culture conditions ... 30
   2.1.2. Isolation of plasmid DNA from bacteria ... 31
   2.1.3. Restriction of plasmid DNA ... 31
   2.1.4. Agarose gel electrophoresis of plasmid DNA fragments ... 31
   2.1.5. Isolation of DNA fragments ... 31
   2.1.6. Ligation of plasmid DNA ... 31
   2.1.7. Sequencing ... 32
   2.1.8. Transformation of *E. coli* and *A. tumefaciens* ... 32

2.2. *Agrobacterium*-mediated transformation of barley cv.
'Golden Promise' 33
   2.2.1. Production of donor plants and growth conditions 33
   2.2.2. Plant tissue culture media 34
   2.2.3. Gene transfer to immature embryos 36
   2.2.4. Generation of transgenic plants 37
   2.2.5. Analysis of transgenic plants 38
      2.2.5.1. Isolation of genomic DNA from plant material 38
      2.2.5.2. PCR 39
      2.2.5.3. Ploidy level analysis 40
      2.2.5.4. Southern blot 40
         2.2.5.4.1. Blotting of separated barley DNA fragments 41
         2.2.5.4.2. DIG labelling and hybridisation 41
      2.2.5.5. Histochemical detection of *gus* reporter gene expression 41
      2.2.5.6. Microscopic detection of *gfp* expression 42
      2.2.5.7. Leaf assay for hygromycin resistance 42
2.3. Production of doubled haploids from co-transgenic T_0 plants 43
   2.3.1. Plant tissue culture media 43
   2.3.2. Stress treatment of microspores 46
      2.3.2.1. Cold treatment of harvested spikes 46
      2.3.2.2. Stravation treatment of isolated microspores 46
   2.3.3. Embryogenic pollen cultures 46
   2.3.4. Identification of selectable marker-free, transgenic segregants 48
   2.3.5. Colchicine treatment of haploid plants 48
2.4. Analysis of sexually generated line 49
2.5. Statistical treatment 50

3. RESULTS ... 51
3.1. Binary vectors ... 51
3.2. Primary transgenic (T_0) plants ... 52
   3.2.1. Evaluation of sister plants derived from the same embryo ... 56
   3.2.2. Genetic transformation and co-transformation ... 61
   3.2.3. Transgene copy numbers in primary co-transgenic lines ... 66
   3.2.4. Ploidy variation among the primary transgenic regenerants ... 69
3.3. Doubled haploids bred from co-transformant selections ... 71
   3.3.1. Embryogenic pollen cultures ... 72
   3.3.2. Environmental influence on the formation of doubled haploids ... 77
   3.3.3. Analysis of individual doubled haploid plants ... 78
   3.3.4. T-DNA segregation in populations of doubled haploids ... 84
3.4. Sexually generated T1 lines ... 89
3.5. Time frame of the established method ... 90

4. DISCUSSION ... 92
4.1. Efficiency of the established method ... 92
4.2. Integration of recombinant DNA in the barley genome ... 102
4.3. Further characteristics of the immature barley genetic transformation and DH production ... 103
4.4. Identification of factors influencing the DH production efficiency ... 106

5. ACKNOWLEDGMENTS ... 107

6. REFERENCES ... 108

7. APPENDIX ... 118

List of abbreviations

bar	phosphinothricin acetyltransferase
CCM	co-culture medium
CIM	callus induction medium
d35S	doubled enhanced promoter of the cauliflower mosaic virus
DH	doubled haploid
DMSO	dimethylsulfoxid
DNA	deoxyribonucleic acid
GMO	genetically modified organism
GOI	gene of interest
gus	ß-glucuronidase
hpt	hygromycin phosphotransferase
IE	immature embryo
KBP	Kumlehn´s barley pollen medium
KBPD	Kumlehn´s solid barley pollen medium
MCS	multiple cloning site
nos	nopaline synthase
npt	neomycin phosphotransferase
ORI	origin of replication
pat	phosphinothricin acetyltransferase
PCR	polymerase chain reaction
PEG	polyethylene glycol
PRM	plant regeneration medium
SM	selectable marker gene
SMB	starvation medium for barley
T-DNA	transfer DNA
uidA	ß-glucuronidase
X-gluc	5-bromo-4-chloro-3-indolyl glucuronide

Bacteria

E. coli *Escherichia coli*
A. tumefaciens *Agrobacterium tumefaciens*
A. rhizogenes *Agrobacterium rhizogenes*

INTRODUCTION

1. INTRODUCTION

The main goal of the present study was the generation of homozygous selectable marker-free transgenic barley plants via segregation in populations of doubled haploid (DH) plants derived from embryogenic pollen cultures, following co-transfer of selectable marker (SM) and gene-of-interest (GOI) as mediated by *Agrobacterium*. The selectable marker was the *hygromycin phosphotransferase (hpt)* gene, directly coupled with *green fluorescent protein (gfp)* and the model gene-of-interest was *β-glucuronidase (gus)*.

1.1. Agroeconomic importance of barley

Barley *(Hordeum vulgare)*, a member of the grass family *Poaceae*, was domesticated about 10 000 years ago in the Near East. It is ranked as the world's fourth crop in terms of acreage and production, and is probably the oldest crop grown by man. Its grain is used both for animal feed and malting, with minor usage in the health food and bioethanol sectors. The simple genetics displayed by barley has for many years encouraged its exploitation as a genetic model, and more recently this has been extended into the field of transgenesis.

It is one of the most widely distributed crops. Among the temperate cereals it is one of best adapted to low rainfall and poor soil conditions. Barley is a major food resource in some regions of the world, including North Africa, the Near East, central Asia, in the Horn of Africa, in the Andean countries and the Baltic States.

There are two forms of barley spikes, the two-rowed and the six rowed varieties. Six-rowed barley, which evolved from the two-rowed variant through a mutation, is better for dry or short growing season, but for other conditions the two-rowed barley gave better yield. Furthermore the carrying stem of two-rowed barley is stronger with superior grain formation and overall grain count. Despite these facts, the six-rowed form is far more widely cultivated all over the world.

INTRODUCTION

The main research on barley deals with fodder and brewing quality improvement to alter the structural grain constituents (starch, proteins, lipids, cell walls) or the enzyme activities that mobilise storage reserves in the seed. Barley seed is a good resource of two components, which are in the focus of attention: tocols (E vitamin) and β-glucans.

Tocols (tocopherols and tocotrienols) have antioxidant activity (Kamal-Eldin and Appelvist 1996) and reduce serum LDL-cholesterol (Wang et al. 1993).

Barley-derived ß-glucan rich diet has several health benefits, among them reduced rate of sugar absorption, a decrease of postprandial glucose, an attenuated glycemic response and reduced risk of cardiovascular diseases (Goedeke et al. 2007).

$(1\rightarrow3)(1\rightarrow4)$ mixed linked β-glucans (β-glucans) of the non-starch polysaccharide family are the major components of barley endosperm cell walls. Its viscosity enhancing property may cause problem in brewing, to modify malting quality by decreasing the soluble β-glucan content in the wort, the gene 1,4-β-glucanase is transferred into the plant to break down the glucan thus improving the filtration rate of beer. Modified grains can also be fed to animals which have difficulty to digest the long chain glucans in the barley cell walls (Nuutila et al. 1999; Xue et al. 2003).

INTRODUCTION

1.2. Genetic transformation of barley

Various transgenic commodity crops, such as herbicide resistant canola, soya and pest resistant maize, are currently grown on millions of hectars around the world. The first transgenic cultivar arrived on the market approximately 15 years ago. Although efficient transformation methods were established first for dicotyledonous plants, the first stable transformation methods for "non-host" monocotyledonous barley appeared in the middle of the 1990s and since then many reports have been presented by various labs (Carlson et al. 2001; Cho et al. 1998; Fang et al. 2002; Funatsuki et al. 1995; Hensel and Kumlehn 2004; Holm et al. 2000; Holme et al. 2006; Jähne et al. 1994; Kumlehn et al. 2006; Matthews et al. 2001; Murray et al. 2004; Nuutila et al. 1999; Patel et al. 2000; Ritala et al. 1994; Salmenkallio-Marttila et al. 1995a; Stahl et al. 2002; Tingay et al. 1997; Travella et al. 2005; Trifonova et al. 2001; Wan and Lemaux 1994; Wang et al. 2001; Wu et al. 1998; Zhang et al. 1999). Up till now, transformation technology of barley is confined to efficiently regenerating cell culture systems of totipotent cells and tissues. Plant regeneration from differentiated tissues, such as leaf and root cells, is not established so far (reviewed by Goedeke et al. 2007)

1.2.1. Methods of gene transfer

Lazzeri et al. (1991) were the first to report about stable transformation of barley immature embryo derived protoplasts. The PEG-induced DNA uptake resulted stable expression of the introduced genes, but no transgenic regenerants were obtained. Later on plants were produced with a similar method (Funatsuki et al. 1995), but with a very low efficiency.

When it comes to efficient barley transformation technology, the spring cultivar 'Golden Promise' is the model genotype. Biolistic gene transfer to pre-cultured immature embryos and microspore derived embryos has led to the establishment of stable transformation of the cultivar (Wan and Lemaux 1994). Tingay et al. (1997) were the first to conduct *Agrobacterium*-mediated gene transfer to immature barley

INTRODUCTION

embryos. This method served as a basis for the further improvement in *Agrobacterium*-mediated DNA transfer methods (reviewed by Goedeke at al. 2007). Matthews et al. (2001) further improved the protocol of Tingay and co-workers directly co-culturing immature embryos with agrobacteria, and without prior biolistic wounding.

Gene transfer to cultivars other than 'Golden Promise' is also possible, but because of poor reproducibility, efficiency and expression, scientists trying to introduce foreign genes to elite barley cultivars are still facing a challenge. However, a growing number of reports have been published about genetic transformation of cultivars other than 'Golden Promise', albeit often associated with poor regeneration efficiency (Cho et al. 1998; Hensel et al. 2008; Manoharan and Dahleen 2002; Murray et al. 2004; Nobre et al. 2000; Ritala et al. 1994; Trifonova et al. 2001; Wang et al. 2001; Wu et al. 1998).

Kumlehn et al (2006) established a method of *Agrobacterium*-mediated gene transfer to embryogenic pollen, where 1.3-8.9 transgenic Igri plants were obtained per 10^6 cultivated immature pollen.

Holme et al. (2006) used isolated barley ovules for *Agrobacterium tumefaciens*-mediated gene transfer, with transformation efficiency comparable to the methods routinely used.

1.2.2. Vector systems

Several strategies exist for the introduction of foreign genes with the use of binary vector systems. *Agrobacterium tumefaciens* can be used for the transfer of heterologous genes. In these systems T-DNA is located *trans* on a separate replicon, while the disarmed Ti plasmid contains the Vir elements (*vir* helper) necessary for the gene transfer (Hellens et al. 2000). Additional copies of the VirB, VirC1 and VirG genes are responsible for the hypervirulence of the some *Agrobacterium* strains or so called superbinary vectors (Komari et al. 1996).

INTRODUCTION

Table 1-1
Agrobacterium strains and binary vectors used for stable transformation of barley

Strain	Biovar	Ti-plasmid	Opine
AGL-1	Biovar l.	pTiBo542ΔT-DNA	Nopaline
AGL-0	Biovar l.	C58 pTiBo524	Nopaline
LBA4404	Biovar l.	pAL4404	Octopine
GV3101	Biovar l.	pMP90 (pTiC58ΔT-DNA)	Nopaline
EHA101	Biovar l.	pEHA101 (pTiBoΔT-DNA	Nopaline

For the *Agrobacterium*-mediated stable transformation of barley, several binary vectors have been used (Table 1-1), which consist of the T-DNA flanked by the left and right border seqences, multiple cloning sites, origin of replication (ORI) for *E. coli* and *A. tumefaciens*. The ORIs used in binary vectors of different agrobacteria strains are to our knowledge confined to pVS1, RK2 and pSP72. Except for GV3101, which resulted in comparatively low transformation efficiency (Kumlehn et al. 2006), all *Agrobacterium* strains shown in Table 1-1 are hypervirulent.

1.2.3. Integration of T-DNA sequences

The first report about the integration patterns of T-DNA into the barley genome was written by Stahl et al (2002), who expressed human genes in transgenic plant tissue. The genomic and T-DNA junctions were determined with biotinylated primers specific for target sequences of the vector border regions. The flanking genomic DNA was cut with a restriction enzyme and ligated to an adapter. The two DNA strands were separated and PCR was carried out. The integration patterns proved to be similar to that observed in dicotyledonous plants. Linkage groups in one single locus occurred with high frequency (50%) in inverted (head-to-tail) configuration, which means that the right border is always attached to a left border and the vice versa, but never two right or left borders are adjacent to each other (head-to head or

INTRODUCTION

tail-to-tail). Deletions from the border sequence were also observed. Right border region flanking the T-DNA seem to be highly conserved, but on the other hand the left border shows greater variability (Fang et al. 2002).

A study aiming at the comparison between biolistic and *Agrobacterium*-mediated T-DNA transfer revealed a difference in the number of integrated copies (Travella et al. 2005), although the results are not supported with statistical analysis. *Agrobacterium* integrated rather few copies of the transgene with minimal rearrangements in all cases, while 60% of the transgenic barley lines derived from particle bombardment integrated more than eight copies of the transgene with extensive DNA rearrangements and multiple integration events. Furthermore, transgene silencing only occurred in lines obtained by biolistic transformation. The results published by Lange et al. (2006) also support the data on *Agrobacterium*-mediated gene transfer. Genomic DNA of 52% of the regenerants obtained by immature barley embryo transformation contained single copy, 33% had two to three copies, and 15% of the independent lines had four inserions. Furthermore, unprecise DNA integration was determined, deletions and integration of backbone vector pieces might occur. Similarly, Hensel et al. (2008) obtained 50% of the plants with single copy, 30% with two, 10% with three and 9% had more than three transgenic fragments integrated in their genome. Holme et al. (2006) analysed the integration pattern of the transgene in independent regenerants obtained by gene transfer to isolated ovules. 37% of the lines contained single copy, 53% two or three copies and 10% four copies, displaying comparable results to those reported on plants derived from gene transfer to immature embryos.

An important issue is the location of T-DNA in the plant genome. Choi et al. (2002) identified integration loci with three or more copies. No preferred integration site was detected, but a trend was shown towards the distal end of the chromosomes. In a report physical and genetic mapping of T-DNA integration sites in 19 independent transgenic barley lines obtained by particle bombardment of immature embryos was conducted (Salvo-Garrido et al. 2004). A total of 23 transgene integration sites were

INTRODUCTION

detected in five (2H, 3H, 4H, 5H and 6H) of the seven barley chromosomes. The integration pattern shows a rather non-random distribution and the genes are integrated preferably in the gene rich telomeric and subtelomeric regions. In specific regions of the choromosomes 4H and 5H clusters of transgenes were observed

1.2.4. Selectable markers and reporter genes

The most commonly used marker genes are antibiotic or herbicide resistance genes. They make the modified cells able to detoxify substances which would otherwise be fatal to them (Miki and McHugh 2004). Usually, selectable markers have no impact on plant growth or development in the absence of selective conditions. Expression of selectable marker genes enables an efficient production and identification of transgenic cells. They are usually introduced in a linked position to the target gene, which confers the desired new trait, and render transgenic cells, tissues and plants resistant to selective pressure.

The first selectable marker used for stable barley transformation was neomycin phosphotransferase (*nptII*) isolated from *E. coli*, providing transgenic cells with kanamycin resistance (Funatsuki et al. 1995; Lazzeri et al. 1991; Nobre et al. 2000; Ritala et al. 1994; Zhang et al. 1999).

Later on, other selection systems were introduced, two of which have been extensively used in barley; the phosphinothricin acetyltransferase (*bar/pat*) genes from *Streptomyces* species (Cho et al. 1998; Jähne et al. 1994; Kumlehn et al. 2006; Tingay et al. 1997; Wan and Lemaux 1994; Wu et al. 1998) that confer resistance to the herbicide phosphinothricin and its derivatives, and the *E.coli* derived hygromycin phosphotransferase gene (*hpt*) confering resistance to the antibiotic hygromycin B. The *hpt*-system proved to be the most efficient in barley transformation (Cho et al. 1998; Coronado et al. 2005; Hagio et al. 1995; Hensel et al. 2008; Holme et al. 2006; Kumlehn et al. 2006; Manoharan and Dahleen 2002; Matthews et al. 2001; Murray et al. 2004; Wu et al. 1998).

INTRODUCTION

A conditional negative selection system was developed for barley (Koprek et al. 1999), where a naturally non-toxic substrate triggers phytotoxic properties in transgenic cells. Use of a positive selection system does not lead to the death of non-transgenic cells, but provides the transformed ones with beneficial properties, so they can thrive on medium unsuitable for the cells lacking the transgene. This method does not require the use of herbicide or antibiotic.

Reed et al. (2001) successfully introduced the phosphomannose isomerase (*pmi*) gene, which converts mannose-6-phosphate into fructose-6-phosphate, in barley among other plants with 3% transformation frequency. Cells expressing the *pmi* gene thrive on medium containing mannose solely as carbon source, non-transformed cells cannot grow on this medium.

Reporter genes are used to indicate their expression or the expression of the target gene in the target cell, tissue, organ or in the entire organism. They must be integrated into transcriptionally active regions in the genome. Regeneration efficiency of transgenic barley calli expressing different reporter genes might differ (Murray et al. 2004).

The reporter function of the green fluorescent protein (*gfp*) gene (Ahlandsberg et al. 1999; Carlson et al. 2001; Chiu et al. 1996; Fang et al. 2002; Holme et al. 2006; Kumlehn et al. 2006; McCormac et al. 1998; Murray et al. 2004; Wang et al. 2001) and the ß-glucuronidase (uidA, *gus*) gene (Cho et al. 1999; Cho et al. 1998; Hagio et al. 1995; Jähne et al. 1994; Kumlehn et al. 2006; Lazzeri et al. 1991; Manoharan and Dahleen 2002; Murray et al. 2004; Nobre et al. 2000; Ritala et al. 1994; Tingay et al. 1997; Wan and Lemaux 1994; Zhang et al. 1999) are commonly used to monitor gene expression in transgenic barley tissue. The disadvantage of the *gus* reporter system is that it is a destructive technique, thus leading to the death of the tissue sample analysed, in contrast to the application of the green fluorescent protein gene, which encodes a relatively small protein with various available derivatives of different emission spectra.

INTRODUCTION

There are reports on the expression of the firefly (*Photinus pyralis*) luciferase reporter gene in barley after particle bombardment of immature embryos (Harwood et al. 2002; Schledzewski and Mendel 1994).

1.2.5. Generation of marker-free lines

Although selectable markers are desirable for the efficient recovery of transgenic regenerants, they have no further purpose once a transgenic plant has been developed. Moreover, the presence of a selectable marker prevents the use of the same gene for any successive round of transformation using another effector gene. In addition, retaining selectable markers which encode resistances to antibiotics is considered in some quarters to be somehow risky, and so commercially grown transgenic plants are often required to be free of those markers. Since 2002, the EU Deliberate Release Directive, which has been in effect, requires "the phasing out of the use of antibiotic-resistance markers in GMOs which may have a harmful impact on human health or the environment".

Several strategies have been elaborated to remove selectable markers from transgenic plants, while retaining the gene of interest (GOI), such as use of intra-genomic relocation of the transgenes using site-specific recombination systems and transposable elements; homologous recombination; and co-integration of transgenes in an unlinked manner, followed by segregation of the T-DNAs in the T_1 generation (Hohn et al. 2001; Shrawat and Lörz 2006; Yoder and Goldsbrough 1994).

Recombinases are used, among other purposes, for the elimination of undesired DNA sequences. The Cre protein, from P1 bacteriophage, is a site-specific DNA recombinase, which is applied to delete a segment of DNA flanked by *lox* recognition elements in the genome, if the *lox* repeats are in a direct orientation. The selectable marker, flanked by specific recognition sites and integrated together with the gene-of-interest in a transgenic unit, is excised by the enzyme, thus selectable marker-free transgenic plants are obtained. Selectable marker genes were successfully excised from tobacco (Dale and Ow 1991; Gleave et al. 1999; Odell et

INTRODUCTION

al. 1990) *Arabidopsis thaliana* (Russell et al. 1992), rice (Hoa et al. 2002), maize (Zhang et al. 2003), and wheat (Srivastava and Ow 2003) using the Cre/*lox* recombinase system.

Other recombinase systems are used as well for the elimination of unwanted DNA sequences. The FLP/FRT system of the 2 µ plasmid of *Saccharomyces cerevisiae* was used to get rid of the selectable marker in both dicots, such as tobacco and *Arabidopsis thaliana* (Kilby et al. 1995), and monocots, e.g maize (Lyznik et al. 1996).

The elimination of the selectable marker using the R-RS system of the pSR1 plasmid of *Zygosaccharomyces rouxii* was established for *Arabidopsis thaliana* (Onouchi et al. 1995), tobacco (Sugita et al. 2000) and rice (Endo et al. 2002).

An irreversible site-specific recombination system is the integrase-*att* from *Streptomyces* phage phiC31 (Thorpe and Smith 1998), which is used for the elimination of selectable markers (Kittiwongwattana et al. 2007; Ow 2007) from transgenic plants. Unlike, Cre/*lox* and FLP/FRT systems, the phiC31 integrase (that mediates recombination between bacterial *attB* and phage *attP* attachment sites) alone cannot reverse the recombination reaction.

The recombinase is not necesseraly included in the T-DNA unit, it might also later be delivered into the plant by secondary transformation (Dale and Ow 1991; Lyznik et al. 1996; Odell et al. 1990), transient expression (Gleave et al. 1999; Jia et al. 2006; Kopertekh et al. 2004) or sexual crossing with a plant expressing the protein (Bayley et al. 1992; Kerbach et al. 2005; Kilby et al. 1995; Russell et al. 1992). The transgenic T-DNA sequence to be evicted and the gene-of-interest, originally coupled to each other in one transgenic unit, are thus separated from each other.

Transposases are also suitable for the production of selectable marker-free transgenic plants, such as the maize *Ac/Ds* elements, which consist of two essential components, the transposase coding gene (*Ac*) and the inverted repeat termini (*Ds*). The transgenes are incorporated within *Ds* elements and its maize transposon in the genomic DNA. *Ds* elements are stable in the absence of *Ac* and lack transposase function. Transgenic

INTRODUCTION

sequences integrated between the *Ds* elements can be mobilised to new genomic locations in the presence of the *Ac* transposase gene (Kunze 1996; Lassner et al. 1989; Masterson et al. 1989). The advantage of this is that after the relocation of the transgene to new chromosomal region, altered expression level might occur, caused by "position effect" (Yoder and Goldsbrough 1994).

Transposable elements retain their transposition competence when introduced in other plant species. The selectable marker was removed from tobacco, aspen (Baker et al. 1986; Ebinuma et al. 1997), tomato (Goldsbrough et al. 1993) and rice (Cotsaftis et al. 2002), with the use of transposable elements, where the excision does not necessarily lead to their reintegration (Belzile et al. 1989; Gorbunova and Levy 2000).

The occurrence of a DNA deletion in tobacco is described through intra-chromosomal recombination between two homologous regions (Puchta 2000; Zubko et al. 2000). However, up till now this system is not eligible for the efficient elimination of the selectable marker.

Another strategy involves the introduction of the selectable marker and the GOI on separate T-DNA sequences (co-transformation), and relies on their integration sites being different. Selectable marker-free plants retaining the GOI can then be selected, provided that the two transgenes are not linked in *cis* (Coronado et al. 2005; De Block and Debrouwer 1991; Depicker et al. 1985; Komari et al. 1996; Matthews et al. 2001; McKnight et al. 1987). A co-transformation experiment of this type in rice and tomato has been reported by Komari et al. (1996), in which two T-DNAs were carried by a single plasmid, but separated from one another by some 15 kb. Its outcome was that over half of the regenerants were selectable marker-free but GOI positive. A similar experiment in barley, involving two T-DNAs separated from one another by only a short spacer, produced a co-transformation frequency of 66%; the GOI was separable from the selectable marker in the progeny of about a quarter of the co-transformants (Matthews et al. 2001).

INTRODUCTION

Agrobacterium-mediated co-transformation is preferred, because it leads to unlinked T-DNA integration events with higher probability than particle gun mediated bombardment of the transgenes. Different ways of *Agrobacterium* mediated T-DNA transfer exist to obtain co-transgenic plants: the use of a mixture of strains (mixture methods) or delivery of T-DNAs from a single strain (single-strain methods). The two plasmids/one strain method means mixture of the same *Agrobacterium* strain (e.g LBA4404) harbouring plasmids with different T-DNAs (De Block and Debrouwer 1991; Komari et al. 1996; McKnight et al. 1987). Coronado et al. (2005) mixed two different *Agrobacterium tumefaciens* strains (LBA4404 and AGL-1) to obtain co-transgenic barley cv. 'Golden Promise' (two plasmids in two different strains method), giving rise to selectable marker free, homozygous T_1 progeny. Single strain methods, with the introduction of two transformation plasmids in one *Agrobacterium tumefaciens* clone (Daley et al. 1998; De Framond et al. 1986; Komari et al. 1996), and use of a binary plasmid, which contains two T-DNAs (Komari et al. 1996; Matthews et al. 2001; Stahl et al. 2002) are also applied for co-transformation experiments.

Finally, some attempts have been made to avoid the use of selectable markers altogether, instead relying on a specific phenotype associated with the expression of the GOI itself (Erikson et al. 2004; Reed et al. 2001). Holme et al. (2006) successfully obtained 0.8 stable transgenic barley plants per 100 isolated barley ovules, without use of any selective conditions.

1.3. Haploid technology

Haploid technology uses haploid cells to produce plants via callus or embryo formation. The originally haploid genome of the regenerants could be doubled either autonomously or by chemical treatment to obtain instantly homozygous plants. For the release of new cultivars this method is routinely applied to accrelerate the breeding procedure (Pickering and Devaux 1992). The particular value of this technology lies in the fact that every indiviual DH-line is a product of random

INTRODUCTION

meiotic recombination, but identically reproducible. Moreover, the technique is also widespread to produce mapping populations used in basic and applied research.

1.3.1. Generation of doubled haploid barley

The major techniques to produce DH barley are anther (Clapham 1973) and ovary culture (Dunwell 1985), interspecific hybridization with *Hordeum bulbosum* L., the use of the haploid initiator gene (Hagberg and Hagberg 1980; Kasha and Reinbergs 1982), and the culture of immature pollen rendered competent to undergo embryogenic development. Embryogenic pollen cultures offer great potential for the generation of DH populations. The first report about induced callus formation in barley pollen cultures appeared in 1982 (Sunderland and Xu 1982), and by now several reports exist on successful production of immature pollen derived plants (Hoekstra et al. 1992; Hunter 1987; Köhler and Wenzel 1985; Olsen 1991; Ziauddin et al. 1990). Moreover, isolated late uninucleate microspores have been considered a valuable material for genetic transformation as well, either by particle bombardment (Jähne et al. 1994), electroporation of protoplasts of microspore-culture origin (Salmenkallio-Marttila et al. 1995a) or *Agrobacterium*-mediated gene transfer (Kumlehn et al. 2006; Wu et al. 1998).

Much effort was done to improve embryo/callus formation and regeneration efficiency of barley embryogenic pollen cultures (Cistué et al. 1995; Kao 1993; Kasha et al. 2001; Mordhorst and Lörz 1993; Ritala et al. 2001; Salmenkallio-Marttila et al. 1995b; Scott and Lyne 1994), e.g. through phenylacetic acid treatment (Ziauddin et al. 1992), and co-cultivation with ovaries (Li and Devaux 2001).

In barley, several common types of inductive treatments exist in order to provide the signal the microspores need to be switched from the gametophytic to the sporophytic development pathway (Sunderland et al. 1978). The common pre-treatment methods are cold shock to anther cultures (Hunter 1987), spikes (Coronado et al. 2005), isolated microspores (Indrianto et al. 1999; Mejza et al. 1993); starvation (Gustafson

INTRODUCTION

et al. 1995; Kumlehn and Lörz 1999; Olsen 1991; Touraev et al. 1997); heat (Touraev et al. 1996) and gametozide-like substances (Zheng et al. 2001).

In barley the prevailing methods are the application of cold stress (+ 4 °C) to anther cultures, spikes and isolated pollen grains in dark, and nutrient starvation (Coronado et al. 2005; Li and Devaux 2003), followed by co-culture of immature pistils, which is likely to provide the developing embryogenic culture with signal molecules (Coronado et al. 2005; Köhler and Wenzel 1985).

The ability of isolated immature pollen to form embryogenic calli and plantlets is highly cultivar dependent (Li and Devaux 2001; Ziauddin et al. 1990). In order to fulfil the major requirement of producing a sufficient number of plants from embryogenic pollen cultures, the isolation and regeneration protocol must be very efficient.

1.3.2. Simplified segregation of transgenes in populations of doubled haploids

The working hypothesis of the present study is that doubled haploid technology can be used for the rapid and efficient production of selectable marker free transgenic T_1 barley plants. Primary plants containing the selectable marker and the gene of interest are obtained from co-transformation. If the T-DNAs are integrated in the plant genome in an unlinked manner, they segregate during the meiotic phase of pollen formation. Isolated microspore cultures of such segregating lines produce selectable marker-free transgenic doubled haploid progeny plants, which are easily identified among individuals of the relatively small T_1 population, without further need of segregation analysis of their offspring. Moreover, the desired plants are instantly homozygous for the transgene, which is of great benefit for scientific and breeding purposes.

INTRODUCTION

1.4. Scope of the thesis

Primary transgenic barley plants (T_0) were generated via *Agrobacterium*-mediated gene transfer to immature embryos using separate T-DNAs, one for the selectable marker (SM) gene hygromycin phosphotransferase *(hpt)*, directly coupled with a green fluorescent protein gene *(gfp)* to be used as additional screenable marker, and the other for the model gene of interest (GOI) ß-glucuronidase *(gus)*, without any selectable marker coupled. Different *Agrobacterium* strain/ binary vector combinations were compared to ultimately identify the most efficient way of uncoupled co-integration of the T-DNAs (Figure 1-1 A). To facilitate the generation of homozygous transgenic SM-free lines in a novel approach, co-transformation was combined with haploid technology. Uncoupled T-DNAs present at hemizygous state in primary transformants are randomly and independently distributed to the pollen grains during male meiosis (Figure 1-1 B). Thus, homozygous transgenic selectable marker-free plants can be instantaneously produced and identified amongst doubled haploid (DH) plants generated from embryogenic cultures of segregating pollen populations (Figure 1-1 C-D).

INTRODUCTION

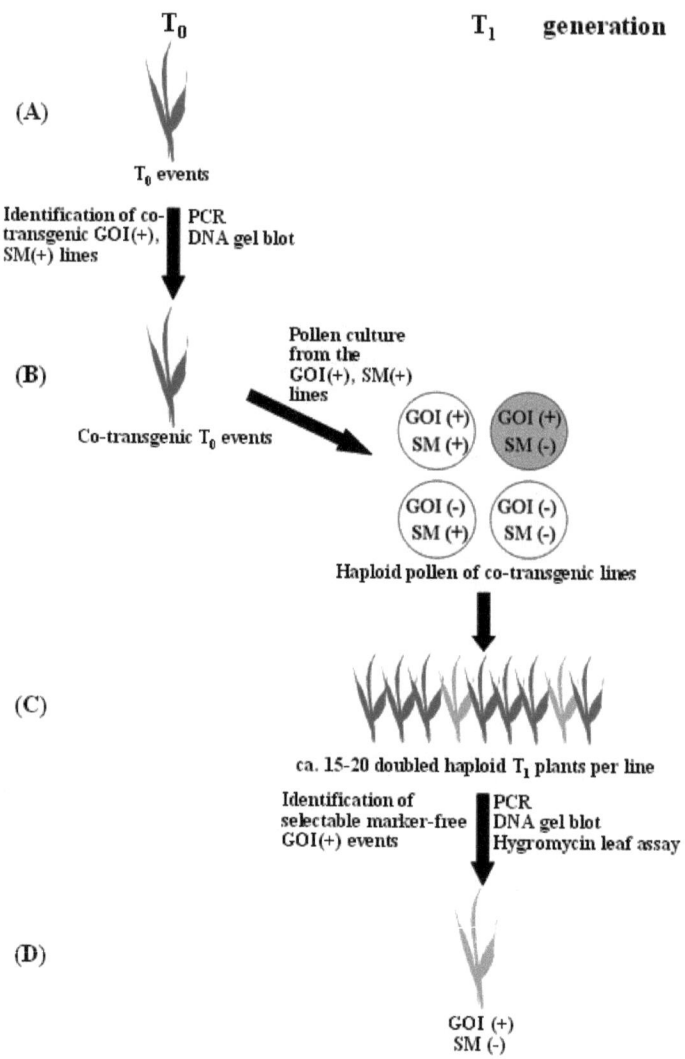

Figure 1-1
Schema for the production of selectable marker-free transgenic barley. (A) Immature embryos were used as the explant subjected to inoculation with *Agrobacterium*. (B) The selectable marker (SM) and the model gene-of-interest (GOI) were co-transformed using separate T-DNAs. (C) Homozygous selectable marker-free GOI positive doubled haploid barley plants were regenerated from embryogenic pollen cultures. (D) If the two T-DNAs are inserted in different chromosomal locations, selectable marker-free GOI positive derivatives can be identified within the doubled haploid progeny.

2. MATERIALS AND METHODS

Agrobacterium-mediated transformation of barley is not genotype-independent. The first stable transformation system of immature barley embryos developed for the model cultivar 'Golden Promise' was established by Tingay et al. 1997. This two kernel row barley with diploid genome (2n=14 chromosomes) is a gamma-ray induced semi-dwarf mutant of the cultivar 'Maythorpe'. It contains the recessive mutation *GPert*, which results short stiff straw and reduced awn length, and also has pleiotropic effects on yield and grain size. The cultivar shows considerable salt tolerance (Forster et al. 1994), but is susceptible to powdery mildew (*Erysiphe graminis* sp. *hordei*).

'Golden Promise' was used in Northern Britain and Scotland as malting barley from the late 1960s to the 1980s. It was favoured by maltsters and distillers, who used two-row barley for malt whisky, because of the unique combination of highly desirable agronomic characters such as earliness (ripening time in August), short stiff straw, easy combining ability, good resistance to grain and ear loss and good malting quality. However, 'Golden Promise' became neglected as many farmers moved to other strains of barley that provided them with higher yield.

2.1. Bacterial strains and vectors

Two *Agrobacterium* strains, LBA4404 and AGL-1 (Table 2-1), and four binary vectors were used in strain mixture (two binary plasmids in two clones of the same *Agrobacterium* strain and two plasmids in two different *Agrobacterium* strains methods) and single strain (two plasmids in one *Agrobacterium* clone, and one Twin-vector harboring two T-DNAs in one *Agrobacterium* clone) co-transformation methods. A total of 14 different variants were applied for immature embryo (gene transfer target) inoculation, Table 2-3.

The cloning steps were conducted in DH5α and DH10B strains of *Escherichia coli* (Sambrook et al. 1989), Table 2-1.

MATERIALS AND METHODS

Table 2-1
Bacterial strains used for cloning and gene transfer to immature barley embryos

Strain	Genotypic specification	Reference
E. coli DH5α	deoR, endA1, gyrA96, hsdR17 (r_k^-, m_k^+), recA1, relA1, λ^- supE44, thi-1, Δ(lacZYA-argFV169), Φ80 Δ lacZ Δ M15, F$^-$	Sambrook et al. 1989
E. coli DH10B	endA1, recA1, galE15, galK16, nupG rpsL ΔlacX74, Φ80lacZΔM15 araD139 Δ(ara,leu)7697 mcrA Δ(mrr-hsdRMS-mcrBC) λ^-, F$^-$	Sambrook et al. 1989
A. tumefaciens LBA4404	AGL0 recA::bla pTiBo542ΔT Mop+CbR	Lazo et al. 1991
A. tumefaciens AGL-1	Ach5 pTiAch5ΔT	Hellens et al. 2000

Plasmids, used for cloning purposes to produce binary vectors applied for the *Agrobacterium*-mediated barley transformation, are presented in Table 2-2.

Table 2-2
Plasmids used for cloning

Vector	Bacterial marker	Origin of replication
pSB227	Sm/Sp^r	pVS1
pGUSi-AB-M	Amp^r	ColE1
pd35S-Nos-AB-M	Amp^r	ColE1
p6U	Sm/Sp^r	pVS1

The pSB227 plasmid (provided by Sylvia Broeders, a former member of the Plant Reproductive Biology Group, IPK-Gatersleben) incorporates hygromycin phosphotransferase (*hpt*) as a selectable marker gene driven by the maize *ubiquitin1* promoter, fused to the *gfp*S65T coding sequence (Chiu et al. 1996) driven by the rice *actin1* promoter (McElroy et al. 1990). The pSB227 plasmid is designated as p*hpt::gfp* to highlight its relevant elements (Figure 2-1).

The binary vector containing the gene-of-interest, without plant selectable marker (p*gus*) was obtained by removing the *hpt* expression cassette from the p6U vector (DNA Cloning Service, Hamburg, Germany), see Figure 2-2. Then the *E. coli* ß-

MATERIALS AND METHODS

glucuronidase gene including the StLS1 intron (Vancanneyt et al. 1990) was cut out from the pGUSi-AB-M (DNA Cloning Service, Hamburg) vector using restriction enzymes SalI and NotI and inserted in the pd35S-Nos-AB-M backbone vector containing the cauliflower mosaic virus (CaMV) doubled enhanced *35S* (d35S) promoter (Odell et al. 1985). Finally, the d35S promoter-*gus* cassette was inserted in the *hpt*-free p6U binary vector with the help of SfiI restriction enzyme.

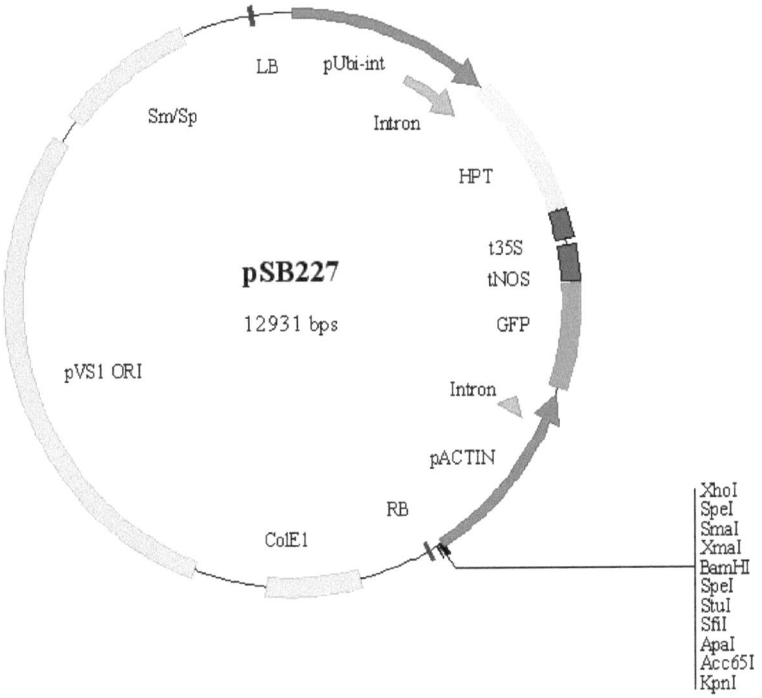

Figure 2-1
Binary vector pSB227 (p*hpt::gfp*) containing the selectable marker gene (*hpt*) coupled with the *gfp* reporter gene

MATERIALS AND METHODS

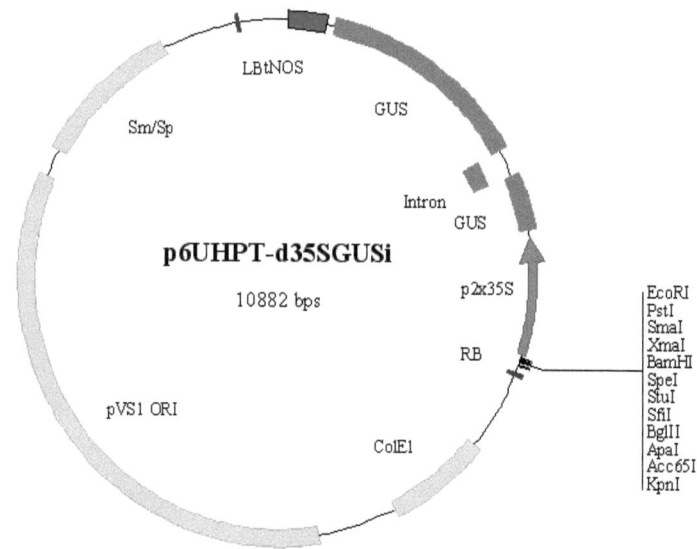

Figure 2-2
Binary vector containing the gene-of-interest (p*gus*) without the selectable marker gene.

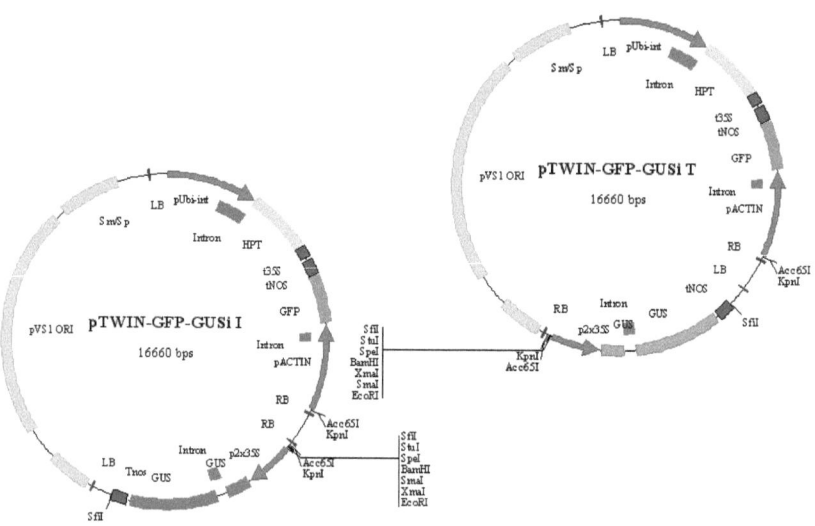

Figure 2-3
Twin vectors. The letters T (tandem) and I (inverted) refer to the different orientations of the *gus* gene related to the *gfp* gene within the vector.

MATERIALS AND METHODS

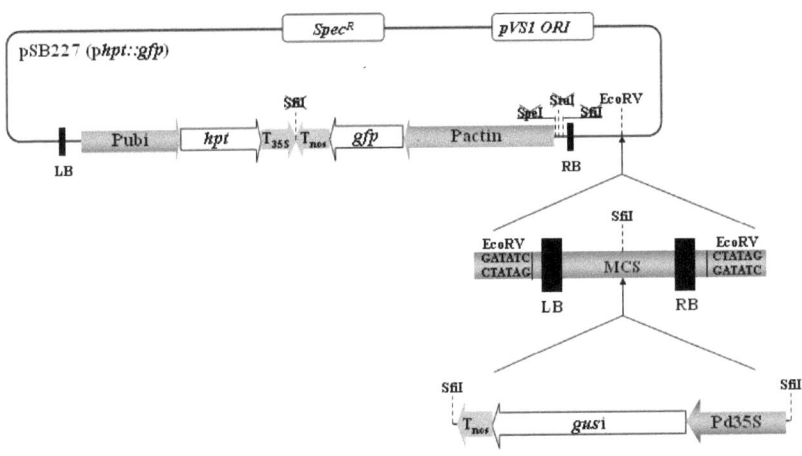

Figure 2-4
The Twin binary vector pair containing two T-DNAs (*gus* and *hpt::gfp*) within a single plasmid. Pubi: maize *ubiquitin1* promoter; Pd35S: doubled enhanced CaMV 35S promoter; Pactin: rice *actin* promoter; Tnos: nopaline synthase terminator; T35S: CaMV 35S terminator; *gfp*: green fluorescent protein coding region; *gus*i: β-glucuronidase (*gusA*) coding region; *hpt*: hygromycin B phosphotransferase coding region; LB: left border; RB: right border; MCS: multicloning site; pVS1 ORI: *E. coli* origin of replication; *Spec*R: coding region for adenylyltransferase conferring bacterial resistance to spectinomycin.

The Twin binary vectors harbour both T-DNAs separated by left and right border sequences (Figure 2-3). Their generation began with the modification of the pSB227 (p*hpt::gfp*) by digestion with *Spe*I and *Stu*I, followed by a 5-3' exonuclease treatment and religation (Figure 2-4). This step also eliminated the *Sfi*I restriction site adjacent to the rice actin promoter sequence, because it overlaps the StuI site. The second *Sfi*I restricion site between the *35S* and *nos* terminator sequences was then removed by *Sfi*I digestion, followed by a 3-5' exonuclease treatment and religation. The Left Border-Multicloning Site-Right Border (LB-MCS-RB) fragment was PCR amplified by primers which incorporated flanking *Eco*RV restriction sites (5'-TAGATATCTGCAAGCTCCACCGGGTGCAAAGCGGCAGC and 5'-CCGATATCATATCCGATTATTCTAATAAACGCTC) using the *hpt*-free p6U vector as template. The LB-MCS-RB fragment was then inserted into the modified pSB227 plasmid at the *Eco*RV site with the help of a TOPO-Cloning kit (Invitrogen) in both possible orientations. The *d35S::gus* sequence was released from the p*gus* vector containing the d35S::*gus*::Tnos cassette by restriction with *Sfi*I and inserted into the pSB227 vector containing the multiple cloning site fragment flanked by the border

MATERIALS AND METHODS

sequences. This resulted in two types of pTwin I and pTwin T binary vectors, differing in their orientation of *gus* in relation to *gfp*; one of these is referred to as T (tandem) and the other as I (inverted).

Binary vectors were transformed into *Agrobacterium tumefaciens* strains AGL-1 and LBA4404 (Hellens et al. 2000; Lazo et al. 1991) by electroporation (see "Transformation of *E. coli* and *A. tumefaciens*") LBA4404 contains additionally the acceptor vector pSB1 (Japan Tobacco Inc., Higashibara, Japan), which confers hypervirulence with the help of accessory alleles VirB, VirC and VirG (Komari et al. 1996).

The 14 different combinations and proportions of the derivative clones of the two *Agrobacterium* strains (AGL-1 and LBA4404) applied for co-culture and their binary vectors- grouped in four methods- are summarized in Table 2-3. The different variants were always compared to a control, because of the highly variable amenability of the donor material due to the environmental conditions which cannot be entirely controlled.

MATERIALS AND METHODS

Table 2-3
Agrobacterium- binary vector combinations used in the comparative co-transformation approach

	Method		*Agrobacterium* strain/ vector variant	Mixture proportion
Mixture methods	Two binary plasmids in two clones of the same *Agrobacterium* strain	1	AGL-1/p*hpt:gfp* AGL-1/p*gus*	50 % 50 %
		2	LBA4404/p*hpt:gfp* LBA4404/p*gus*	50 % 50 %
	Two plasmids in two different *Agrobacterium* strains	3	AGL-1/p*hpt:gfp* LBA4404/p*gus*	70 % 30 %
		4	AGL-1/p*hpt:gfp* LBA4404/p*gus*	50 % 50 %
		5	AGL-1/p*hpt:gfp* LBA4404/p*gus*	30 % 70 %
		6	LBA4404/p*hpt:gfp* AGL-1/p*gus*	30 % 70 %
		7	LBA4404/p*hpt:gfp* AGL-1/p*gus*	50 % 50 %
		8	LBA4404/p*hpt:gfp* AGL-1/p*gus*	70 % 30 %
Single strain methods	Two plasmids in one *Agrobacterium* clone	9	AGL-1/p*hpt:gfp* + p*gus*	-
		10	LBA4404/p*hpt:gfp* + p*gus*	-
	Two T-DNAs in one binary vector in AGL-1	11	AGL-1/pTwin T	-
		12	AGL-1/pTwin I	-
	Two T-DNAs in one binary vector in LBA4404	13	LBA4404/pTwin T	-
		14	LBA4404/pTwin I	-

MATERIALS AND METHODS

2.1.1. Bacterial media and culture conditions

Bacterial culture media, which were prepared either in liquid and solid form, were autoclaved at 120 °C for 20 minutes. Solid media were prepared with 0.8% agar.

E. coli strains were grown on LB-Medium (Silhavy et al. 1984).

MG/L culture medium (Garfinkel and Nester 1980) was used for the growth of AGL-1, and CPY for strain LBA4404 (Komari et al. 1996), see Table 2-4.

Table 2-4
Culture medium applied for the growth of *Agrobacterium* strains AGL-1 and LBA4404

Components		MG/L	CPY
sugars g/l	mannitol	5	-
	sucrose	-	5
amino acid (g/l)	L-glutamic acid	1	-
macro elements (mg/l)	KH_2PO_4	250	-
	NaCl	100	-
	$MgSO_4·7H_2O$	100	500
vitamin (µg/l)	biotin	1	-
miscellanous (g/l)	peptone	-	5
	tryptone	5	-
	yeast extract	2.5	1
pH		7.0	7.2

Antibiotics were added after autoclaving in following concentrations: carbenicillin 100 µg/ml, rifampicin 50 µg/ml, spectinomycin 100 µg/ml, and tetracycline 10 µg/ml.

E. coli strains were grown at 37 °C, while *A. tumefaciens* strains at 28 °C. Liquid media were shaken at 180 rpm in Erlenmeyer flask.

Cryostock cultures of *A. tumefaciens* strains were prepared with 7% glycerol and stored at -80 °C. For the co-culture with immature embryos the content of the tube

MATERIALS AND METHODS

was thawed and put in 10 ml medium without antibiotics and shaken (180 rpm) at 28 °C for about 24 hours, optical density (OD) was set at 0.2-0.25.

2.1.2. Isolation of plasmid DNA from bacteria

Bacterial plasmid DNA was isolated from 2 ml liquid culture, using Qiagen Spin Miniprep Kit (Quiagen/Germany).

2.1.3. Restriction of plasmid DNA

Restriction enzymes from Fermentas were used with appropriate reagent buffer. For cloning purpose, 15-20 µl restriction digest reactions contained 2-4 U endonuclease at incubation temperature given by the producer. Successful Southern blot analysis required the digest of greater amount of DNA (10-30 µg) with 10-20 U enzyme in 50 µl end volume.

2.1.4. Agarose gel electrophoresis of plasmid DNA fragments

DNA fragments were run in 1.2 % agarose gel containing ethidium-bromide (0.3 µg/ml), in order to make DNA visible under UV light. Gel electrophoresis was done in 0.5xTBE buffer (45 mM Tris, 45 mM Borsäure, 1 mM EDTA).

2.1.5. Isolation of DNA fragments

DNA fragments were isolated using Qiagen Gel Extraction Kit (Qiagen/Germany).

2.1.6. Ligation of plasmid DNA

Ligation of DNA with overhanging ends was carried out by Sambrook et al. (1989) using ligase from Fermentas, blunt ends were treated with phosphatase (Antarctic phosphatase, Biolabs New England) in order to prevent self-ligation.

MATERIALS AND METHODS

2.1.7. Sequencing

Sequencing was carried out by AGOWA genomics services according to their protocol.

2.1.8. Transformation of *E. coli* and *A. tumefaciens*

For the production of electrocompetent cells 800 ml appropiate medium was inoculated with 0.8 ml fresh culture of *E. coli* or *A. tumefaciens* strains AGL-1, LBA4404. The liquid culture was shaken (37 °C *E. coli*, 28 °C *A. tumefaciens*, 200 rpm) for 4-5 hours until it reached 0.5-0.8 OD_{600} value. This was followed by a centrifugation (5000 rpm) step at 4 °C. The cells were washed with 800 ml ice-cold distilled water and then centrifuged for 20 minutes at 4500 rpm, this was followed by an other washing step in 400 ml ice-cold distilled water. The pellet was washed with 25 ml 10 % glycerine and centrifuged (10 minutes, 4000 rpm). The cells were resuspended in 1.5-3.0 ml 15 % glycerine and 50 µl aliquots were stored at -80 °C (storage life 6-12 months) in 1.5 ml tubes.

Transformation of bacterial cells was conducted in an electroporation cuvette, containing an aliquot thawed on ice with 2 µl added DNA solution. After electroporation (4-5 msec, 25 µF, 2,5 kV, 200 Ohm; BioRad/USA) 1 ml SOC medium (Table 2-5) was added quickly, then *E. coli* cells were incubated and shaken (180 rpm) at 37 °C for one hour, *A. tumefaciens* cells at 28 °C for 3 hours. Transformed bacteria were plated (100-200 µl) on solid medium containing antibiotics and grown (*E.coli* overnight, *A. tumefaciens* 2 days).

MATERIALS AND METHODS

Table 2-5
Composition of SOC medium

SOC medium components	
bacto-trypton	2 %
yeast extract	0.5 %
KCl	2.5 mM
$MgCl_2$	10 mM
$MgSO_4$	10 mM
glucose (added after autoclaving)	20 mM

2.2. *Agrobacterium*-mediated transformation of barley cv. 'Golden Promise'

The constructs were transformed into barley cv. 'Golden Promise' according to Tingay et al. (1997), but without prior biolisctic wounding. Hygromycin selection was used instead of bialaphos, due to increased efficiency of the previous system. The further improved immature barley embryo transformation protocol presented in this work is based on the method established by G. Hensel and J. Kumlehn, 2004.

2.2.1. Production of donor plants and growth conditions

Seeds of spring barley (*Hordeum vulgare*) cultivar 'Golden Promise' were germinated in substrate mix (Specialmischung Petuniensubstrat) in growth chamber under controlled condition (14/12 °C day/night, 12h light, 20 000 lux, relative humidity ca. 80%) for 10-12 weeks. The plants were fertilised at the beginning of tillering with Osmocote (40g/7.5 l), a long-term fertilizer, containing 19% Nitrogen, 6% Phosphorus and 15% Potassium. In the period of every two weeks the plants were watered with 0.3% Hakaphos Blau (Compo, Germany), a general fertilizer containing 15% Nitrogen, 10% Phosphorus and 15% Potassium, until the stems started to elongate. The plants were placed in a greenhouse cabin (18/14 °C day/light, min. 25 000 lux for 16 h) immediately after the spikes emerged from the leaf sheath.

MATERIALS AND METHODS

Donor plant conditions highly influence the outcome of the experiment (Kasha et al. 1989; Kuhlmann and Foroughi-Wehr 1989).

2.2.2. Plant tissue culture media

Tissue culture media protocols (Table 2-6), used for *Agrobacterium*-mediated immature barley embryo transformation and regeneration of the primary transgenic plants, are based on the publication by Tingay et al. 1997.

MATERIALS AND METHODS

Table 2-6
Barley immature embryo derived tissue culture media

Components		CCM (Tingay et al. 1997)	CIM (Hensel and Kumlehn 2004)	PRM (Hensel and Kumlehn 2004)
Macroelements (mg/l)	NH_4NO_3	1650	1650	320
	KNO_3	1900	1900	3640
	KH_2PO_4	170	170	340
	$CaCl_2 \cdot 2H_2O$	441	441	441
	$MgSO_4 \cdot 7H_2O$	331	331	246
Microelements (mg/l)	H_3BO_3	6.2	6.2	3.10
	$MnSO_4 \cdot 4H_2O$	22.4	22.4	11.20
	$ZnSO_4 \cdot 7H_2O$	8.6	8.6	7.20
	KI	0.83	0.83	0.17
	$Na_2MoO_4 \cdot 2H_2O$	0.25	0.25	0.12
	$CuSO_4 \cdot 5H_2O$	0.025	1.275	0.13
	$CoCl_2 \cdot 6H_2O$	0.025	0.025	0.024
	$Na_2FeEDTA$	36.70	36.70	36.70
Vitamins (mg/l)	B5 vitamins (Duchefa)	-	-	112
	thiamine-HCl	1.0	1.00	10
Amino acids (mg/l)	L-Cysteine	800	-	-
	L-Glutamine	-	-	146
	L-Proline	690	690	-
Sugars (g/l)	maltose monohydrate (Duchefa)	30	30	36
Growth regulators (mg/l)	DICAMBA	2.50	2.50	-
	6-BAP	-	-	0.225
Miscellaneous (g/l)	acetosyringone	0.098	-	-
	casein Hydrolysate	1.00	1.00	-
	myo-Inositol	0.25	0.25	-
	timentin	-	0.15	0.15
	Phytagel (Sygma, Germany)	-	3.0	3.0
pH		5.8	5.8	5.8

MATERIALS AND METHODS

2.2.3. Gene transfer to immature embryos

The transformation protocol applied to immature embryos and the generation of primary transgenic plants followed that of Hensel and Kumlehn (2009). Two *Agrobacterium tumefaciens* strains were used: a hypervirulent derivative of LBA4404 (Komari et al. 1996) and AGL-1 (Lazo et al. 1991). Genetic transformation of barley (*Hordeum vulgare* L.) was carried out using 14 different Agrobacterium/vector combinations, which are specified in Table 2-3. Each of the three replicates making up the entire experiment consisted of the inoculation of 90 immature embryos of the cultivar 'Golden Promise' with each of the 14 combinations. Because it was technically impossible to compare all 14 combinations in a single experimental run, combination 7 (a 1:1 mixture of LBA4404/p*hpt:gfp* and AGL-1/p*gus*) was included as an internal control in each transformation experiment. This 'control' was thus applied in a total of six replicates using 270 embryos each.

All of the below mentioned immature embryo transformation steps were carried out under sterile conditions in a sterile bench.

Developing caryopses of donor barley plants were harvested at around 12 days after pollination for immature embryo isolation and surface sterilized in order to avoid infection of the tissue cultures. Seeds were stirred in 70 % ethanol for 3 minutes, followed by a 20 minutes washing step in 5% NaOCl solution, to which 0.5 ml Tween was added. Finally, seeds were stirred 5 minutes in double distilled autoclaved water, rinsed 5 times and stored, if necessary overnight, at 4 °C.

Immature barley embryos were excised from the caryopses by using forceps and lanzette needle at a stereo microscope under sterile conditions (Tingay et al. 1997). The stadium of the embryos highly determines their capability for genetic transformation, the best are the ones which are transparent in the middle, but white on the side with a diameter of 1.5-2mm. The embryonic axes were dissected and 30 embryos were put in each well of a 6-well-plate (Greiner Bio-One Gmbh, Austria) filled with 2.5 ml co-culture medium (CCM) supplemented with 9.8 mg/l acetosyringone (Table 2-6).

MATERIALS AND METHODS

CCM was removed using a sterile pipet from the plates and 600 µl *Agrobacterium* suspension (OD range 0.2-0.25) was added. The 6 well-plate was vacuum infiltrated for 1 minute at 500 mbar (diaphragm pump MP 201 E from Ilmvac, Ilmenau, Germany) and incubated covered for 10 minutes. The *Agrobacterium* suspension was removed and the embryos were washed with 2.5 ml CCM medium and incubated for another 15 minutes. After another washing step with 2.5 ml CCM medium the plates were transferred to 21 °C in the dark.

2.2.4. Generation of transgenic plants

After 60 hours the embryos, later the derived developing calli, were fortnightly transferred first onto modified Callus Induction Medium (CIM), described by Trifonova et al. 2001, and supplemented with 20, later 50 mg/l hygromycin B (Boehringer, Germany) and 150 mg/l timentin. 10 embryos/ calli were placed on each 9-cm petri dish (Greiner Bio-One Gmbh, Austria) with the scutellum side facing the medium (Hensel et al. 2008). Calli were kept in an incubator at 25 °C in dark. After the induction phase the obtained calli were then transferred onto fresh Plant Regeneration Medium (PRM), supplemented with 25mg/l hygromycin B, in every two weeks (Table 2-7), incubated under light (24 °C, 16/8h light/dark photoperiod, 10 000 lux).

Solid media were prepared from a mixture of the components (Table 2-6), filter sterilized in fourfold-concentration, and the diluted Phytagel (previously solved in double-distilled water and autoclaved for 20 min at 120 °C) used for solidification. During the regeneration process the developing calli produced plantlets.

MATERIALS AND METHODS

Table 2-7
Summary of callus transfer periods, induction and regeneration conditions

	Medium	Time period	Conditions
Co-culture	CCM	60h	21 °C, dark
Callus induction	CIM_{solid} + 20mg/l hygromycin B	2 weeks	24 °C, dark
	CIM_{solid} + 50mg/l hygromycin B	2 weeks	24 °C, dark
Plant regeneration	PRM_{solid} + 25mg/l hygromycin B	2x 4 weeks	24 °C, light

The primary regenerants were transferred into glass tubes (height 100 mm, outer diameter 25mm, Schütt, Germany) containing solid PRM medium supplemented with 25mg/l hygromycin B. After the roots have appeared, the small barley plants were transferred to the greenhouse (14/12 °C day/night, 12h photoperiod, 20 000 lux, relative humidity ca. 80%) in small pots with 6cm diameter. Later on selected lines to be brought to maturation were put in big pots with 16cm diameter. 1 m^2 greenhouse area is required to store 120 small plantlets or 20 mature barley plants.

2.2.5. Analysis of transgenic plants

After the plants were transferred to soil, their genomic DNA was tested for the presence of the gene-of-interest (*gus*) and the selectable marker gene (*hpt::gfp*). Based on the PCR results the co-transformed primary transgenic (T_0) lines containing both *gus* and *hpt::gfp* were selected. Among the positively tested ones further analysis was carried out, ploidy level and copy number of the integrated T-DNAs were determined as well.

2.2.5.1. Isolation of genomic DNA from plant material

Genomic DNA was isolated from leaves of the primary regenerants (Palotta et al. 2000). Fresh leaf material (200-400 mg) was put in a 2 ml Eppendorf tube with two metal beads (diameter 4.1 mm) and stored in liquid nitrogen or at -80 °C until extraction. The frozen leaves were comminuted using a mixer mill for 2 minutes,

MATERIALS AND METHODS

1/27 s (Retsch Mixer mill MM301). 800 µl extraction buffer (1% N-lauryl-sarcosin, 100mM Tris-HCl pH 8.0, 10 mM EDTA pH 8.0, 100 mM NaCl) was added to each tube and vortexed until the clumps were dissolved. 800 µl phenol/chlorophorm/isoamyl alcohol (25:24:1, Roti) was added and suspended by thorough vortexing, which was followed by a centrifugation step at room temperature (5000 rpm, 3 minutes). The upper layer containing dissolved DNA was transferred in clean 1.5 ml Eppendorf tubes. 1/10 vol. (80 µl) 3M Na-acetate (pH 5.2) and 1 vol. (800 µl isopropanol) was added and mixed until white DNA precipitate appeared. After the centrifugation step at 4 °C 13 000 rpm for 10 minutes the supernatant was discarded and the pellet was washed with 1 vol. (800 µl) 70% ethanol. The supernatant was discarded after the final centrifugation step (1 minute, 13 000 rpm) and the pellet was air dryed for ca. 30 minutes. The plant genomic DNA was resolved in 100 µl R40 (40 µg/ml RNase in TE buffer, 10:1 pH 8.0) and incubated at 37 °C for one hour. The samples were stored for shorter time period at 4 °C, longer storage is possible at -20 °C.

2.2.5.2. PCR

Genomic DNA of presumptive transgenic regenerants was used to establish the presence of both *gus* and *hpt::gfp*. PCRs were based on primer pairs specific for either *gfp* or *gus* (Table 2-8). Each PCR involved an initial denaturation step (95°C/5min), followed by 35 cycles of 95°C/30s, 60°C/45s, 72°C/75s, and ending with a final extension step (72°C/7min). PCR products were separated by electrophoresis through 1.2% agarose gels. The length of the *gus* amplicon was 730bp, and that of the *gfp* amplicon was 450bp.

MATERIALS AND METHODS

Table 2-8
Forward and reverse primers used for the identification of the gene-of-interest (*gus*) and the selectable marker (*hpt* coupled with *gfp*) integrated in the genome of barley plants.

Sequence	Name
5'-CCGGTTCGTTGGCAATACTC-3'	GH-GUS F1
5'-CGCAGCGTAATGCTCTACAC-3'	GH-GUS R1
5'-GGTCACGAACTCCAGCAGGA-3'	GH-GFP F1
5'-GACCACATGAAGCAGCACGA-3'	GH-GFP R1

2.2.5.3. Ploidy level analysis

The ploidy level of the primary co-transgenic plants was determined using a flow cytometer (PA1, Partec), precisely measuring the total DNA content of individual nuclei, in order to confirm if the plants are in a haploid, diploid or tetraploid state. This information is especially useful when analysing T_1 populations.

2.2.5.4. Southern blot

DNA gel-blot hybridization (Sambrook et al. 1989) was used to confirm the selectable marker-free status and characterize the segregation pattern of the insertions. The same primers were applied for the production of the *gfp* and *gus* probes as for the PCR. The amplicons were labelled with digoxygenin (PCR DIG Probe Synthesis kit, Roche Diagnostics, Mannheim, Germany) for use as hybridization probes. Ca. 30 µg Genomic DNA was digested with *Hin*dIII (37 °C, overnight) separated by electrophoresis through an 0.8% agarose gel, and transferred to a positively charged nylon membrane (Roche Diagnostics), following the manufacturer's instructions. Each blot was hybridized first with the *gus* probe, then 'stripped' and reprobed with the *hpt* sequence. Hybridization, signal detection and probe stripping were carried out following the DIG Application Guide for Filter Hybridization Manual (Roche Diagnostics, Mannheim, Germany). As negative control isolated genomic DNA from wild type 'Golden Promise' plants was applied. 80 pg of p*hpt::gfp* and p*gus* plasmids served as positive control.

MATERIALS AND METHODS

2.2.5.4.1. Blotting of separated barley DNA fragments

DNA fragments were separated on 0.8% (w/v) agarose gel by electrophoresis at 25V overnight. According to the manufacturers instructions (Roche Diagnostics, Mannheim, Germany) the DNA fragments were transferred to a positively charged nylon membrane (Roche Diagnostics, Mannheim, Germany).

2.2.5.4.2. DIG labelling and hybridisation

DNA fragments, blotted on a positively charged nylon membrane, were hybridized with DIG-dUTP labeled probes, with the use of PCR DIG Probe Synthesis Kit (Roche Diagnostics, Mannheim, Germany). The applied primers were the same as that of used for PCR reactions.

Hybridisation steps, signal visualisation by CDP-Star and detection on chemiluminescent detection film were done according to the DIG Application Guide for Filter Hybridization Manual (Roche Diagnostics, Mannheim, Germany).

2.2.5.5. Histochemical detection of *gus* reporter gene expression

The *gus* reporter gene system is generally used in molecular biology (Jefferson 1987). X-Gluc is the substrate of the ß-glucuronidase enzyme, which converts it to glucuronic acid and 5-bromo-4-chloro-indoxyl. The latter is then oxidised to 5,5'-dibromo-4,4'-dichloro-indigo, a blue coloured product. GUS histochemical staining was applied to embryogenic callus and leaf tissue. The plant material was vacuum-infiltrated (ILMVAC, Laboratory Vacuum System, LVS 301 Zp, Ilmenau, Germany), then held overnight at 37°C in in X-Gluc (5-bromo-4-chloro-3-indonyl-D-glucuronide) solution (Table 2-9). When testing leaves, the chloroplasts, which disturb the blue tone, were extracted by alcohol (60 °C, 2 hours in water bath).

MATERIALS AND METHODS

Table 2-9
Composition of X-Gluc solution

Components	Final concentration	Comment
X-Gluc	1 mg/ml	
methanol	20%	dissolve X-Gluc in methanol
0.5 M Na_2HPO_4/NaH_2PO_4 solution	100 mM	
0.5 M NaEDTA	10 mM	
Triton X-100	0.1%	
K-hexacyanoferrat (II)	1.4 mM	
K-hexacyanoferrat (III)	1.4 mM	set pH at 6.2-7.2 store at -20 °C

2.2.5.6. Microscopic detection of *gfp* expression

GFP expression was screened in callus tissue and root tips, using a Leica MZFLIII fluorescence microscope with a filter set for GFP Plant (Leica Microsystems, Wetzlar, Germany). Reporter gene expression is shown in Figure 3-2 C.

2.2.5.7. Leaf assay for hygromycin resistance

A rapid assay for the detection of *hpt* and *bar* marker genes in transgenic barley leaves has been established (Wang and Waterhouse 1997). Harvested leaves were sprayed with 70% ethanol and stabbed in a 9 cm Petri dish containing PRM medium (Hensel and Kumlehn 2004) supplemented with 200 mg/l hygromycin B. The dish was kept in a light chamber (24 °C, 16h photoperiod, 20 000 lux) for one week. Leaves of plants free of *hpt* bleached on selective medium, while *hpt*-transgenics stay green over one week (Figure 3-9).

MATERIALS AND METHODS

2.3. Production of doubled haploids from co-transgenic T_0 plants

Microspores at the highly vacuolated, pre-mitotic stage were isolated following Coronado et al. (2005). Other experimental details are as given by Kumlehn et al. (2006).

2.3.1. Plant tissue culture media

Descriptions of media prepared for barley microspore starvation treatment (Table 2-10) and embryogenic callus formation (Table 2-11) are based on the publication of Kumlehn et al. (2006). PRM medium (see 2.2.2.) was used for the regeneration of doubled haploid barley plants from microspore derived calli.

MATERIALS AND METHODS

Table 2-10
Composition of microspore starvation medium (SMB)

Components		Final concentration
macroelements (mg/l)	NH_4Cl	53.40
	$CaCl_2$	110.80
microelements (mg/l)	$MnSO_4 \cdot H_2O$	5.25
	H_3BO_3	3.10
	$ZnSO_4 \cdot 7H_2O$	7.20
	$Na_2MoO_4 \cdot H_2O$	0.123
	$CuSO_4 \cdot 5H_2O$	0.025
	$CoCl_2 \cdot 6H_2O$	0.020
	KI	0.166
sugar (g/l)	maltose (Sigma)	144
growth regulator (mg/l)	BAP	0.9
antibiotic (mg/l)	cefotaxime	250
miscellaneous (mg/l)	MES	426
	pH	5.5

MATERIALS AND METHODS

Table 2-11
Composition of KBP liquid and KBPD solid media

Components		Final concentration
macroelements (mg/l)	NH_4NO_3	80
	KNO_3	101
	KH_2PO_4	136
	$CaCl_2 \cdot 2H_2O$	110.8
	$MgSO_4 \cdot 7H_2O$	246.3
microelements (mg/l)	$MnSO_4 \cdot H_2O$	5.25
	H_3BO_3	3.10
	$ZnSO_4 \cdot 7H_2O$	7.20
	$Na_2MoO_4 \cdot H_2O$	0.123
	$CuSO_4 \cdot 5H_2O$	0.025
	$CoCl_2 \cdot 6H_2O$	0.020
	KI	0.166
	NaFeEDTA	27.5
sugar (g/l)	maltose (Sigma)	90
amino acid (mg/l)	glutamine	438
growth regulator (mg/l)	BAP	0.9 (KBP)
		0.225 (KBPD)
vitamin	Kao and Michayluk vitamin solution (Sigma)	1x
antibiotic (mg/l)	cefotaxime	250 (KBP)
	timentine	150 (KBPD)
miscellaneous (g/l)	Phytagel (Sygma, Germany)	3.0 (KBPD)
	pH	5.9

MATERIALS AND METHODS

2.3.2. Stress treatment of microspores

2.3.2.1. Cold treatment of harvested spikes

Spikes, 6 cm from the first knot and covered by the leaf sheath, were harvested from each plant separately. The spikes were surface sterilized with 70% ethanol, the awns were removed and pre-treated at 4 °C for 3-4 weeks in 9-cm Petri dishes kept humid by wet filter paper (Hunter 1987; Mordhorst and Lörz 1993). Five-six spikes were placed in each dish, which was sealed by Parafilm (Pechiney Plastic Packaging, Menasha, WI 54952, USA). The following steps were carried out under aseptic conditions in a sterile bench.

2.3.2.2. Stravation treatment of isolated microspores

SMB liquid medium (Table 2-10) was used for starvation stress of the freshly isolated microspores. The optimal time span of the stress treatment is two days.

2.3.3. Embryogenic pollen cultures

Isolation of microspores in the late uninucleate stage from co-transgenic To 'Golden Promise' barley plants was done according to the protocol established by Coronado et al. 2005.

For the isolation of the microspores 10-15 spikes were cut to ca. 1cm long pieces and put in a sterile, pre-cooled Micro Container. Twenty milliliters of ice cold 0.4 M mannitol (Duchefa) was added and blended using a Waring Blendor power unit (Eberbach, Ann Arbor, MI, USA) for 2x 10 seconds at "low" speed. The suspension was poured into a vessel on ice through a 100 µm nylon sieve (Wilson, Nottingham, UK) and rinsed with 10 ml 0.4M mannitol. The rest was pressed by a sterile forceps to obtain more suspension and put back in the blender. Ten milliliters 0.4M mannitol was added and homogenised for another 2x 10 seconds. The suspension was poured onto the 100 µm sieve and the Micro Container rinsed again with 10 ml 0.4M mannitol. The rest was gently pressed by a sterile forceps and then removed together

MATERIALS AND METHODS

with the sieve. The suspension was poured in a 50 ml srew-cap tube. Five milliliters 0.4M mannitol was taken to rinse the vessel and poured in the srew-cap tube. The suspension was centrifuged at 4 °C, 705 rpm. The supernatant was removed with a sterile 10 ml pipet and using a new sterile pipet the pellet was resuspended in 5 ml ice-cold 0.55M maltose (Sigma) and transferred into a 12 ml tube (Greiner Bio-One Gmbh, Austria). The 50 ml srew-cap tube was rinsed with 1.5 ml 0.4M mannitol and carefully overlayed on the suspension in the 12 ml tube. After the next gradient-centrifugation step (4 °C, 705 rpm) the interphase containing the viable microspores was removed and resuspended in 0.4 ml mannitol in a new 50 ml srew-cap tube setting the final volume at 20 ml. These cells are in a vacuolated mononucleous stadium, able for cell division and production of doubled haploid regenerants. Debris of dead cells is found in the pellet. Twenty microliters of the suspension was pipetted on a haemocytometer in order to count the number of microspores. The supernatant was removed after the following centrifugation step, the pellet was dried by a sterile 1 ml pipet tip, which was pressed to the bottom of the srew-cap tube and the mannitol rest was carefully sucked out without removing the microspore cells. The pellet was resuspended in starvation medium (SMB, Table 2-10) and the concentration set between 100.000- 400.000 microspores/ml aliquoted in 35-mm Petri dishes (Greiner Bio-One Gmbh, Austria) (1 ml culture per dish) and incubated in dark at 21 °C for two days.

After 2 days SMB medium was removed with a 1 ml disposable pipet and KBP medium (Kumlehn's Barley Pollen medium, Table 2-11), including cefotaxim as antibioticum and 5 immature wheat pistils (pre-incubated for one day in KBP medium with a maximum number of 20-30 pieces per 2 ml medium in a 35 mm petri dish.) were added (Hu and Kasha 1997; Köhler and Wenzel 1985; Li and Devaux 2001). The cultures were incubated at 25 °C in the dark. After one week an additional milliliter of fresh KBP (incl. cefotaxim) medium was added, and the cultures were put on a rotary shaker (ca. 50 rpm) at 25 °C in the dark. After two weeks the calli were transferred on filter paper containing solid KBPD medium (Table 2-11) in 9-cm

MATERIALS AND METHODS

dishes (Greiner Bio-One Gmbh, Austria) and incubated at 25 °C in dark. After another two weeks calli were transferred to PRM medium (see Table 2-6, Kumlehn et al., 2006), kept in dark at 25 °C for 5 days and then put in a light chamber (24 °C, 16h photoperiod, 20 000 lux). Calli were transferred to fresh medium in periods of 3x2 weeks. Regenerants were put in sterile boxes containing fresh PRM medium. After the small plantlets developed roots they were transferred to a phytochamber in the greenhouse (14/12 °C day/night, 12h photoperiod, 20 klx, relative humidity ca. 80%).

2.3.4. Identification of selectable marker-free, transgenic segregants

PCR, Southern Blot, *gus* assay and hygromycin leaf test were carried out the same way as explained by "2.2.5. Analysis of transgenic plants".

2.3.5. Colchicine treatment of haploid plants

Colchicine is a toxic chemical, first extracted from Autumn crocus (*Colchicum autumnale*). It is a "mitotic poison", inhibiting mitosis by binding to tubulin, a crucial component. Because colchicine is inhibiting chromosome segregation during cell division, it is applied to induce polyploidy in plant cells, to double the chromosome numbers during cell division (Luckett 1989; Takamura and Miyajima 1996).

Haploid regenerants were diploidized by subjection to colchicine treatment (Thiebaut and Kasha 1978). Selected haploid barley plants, grown in the greenhouse (14/12 °C day/night, 12h photoperiod, 20 klx, relative humidity ca. 80%), were put overnight in a dark chamber at 4 °C before the colchicine treatment, in order to induce mitotic cell division in the plant tissue. Soil was removed from between the roots by washing with tap water. Roots were cut to 3 cm and leaves to 5 cm long pieces. Plants were put in 50 ml screw cap tubes filled with 25 ml colchicine solution (1g/l) and incubated for 6 hours in a light chamber. Plants were then carefully washed, repotted and put in the greenhouse.

MATERIALS AND METHODS

2.4. Analysis of sexually generated lines

When no plants of embryogenic pollen culture origin were obtained, sexual lines of self-pollination were germinated from the seed reserve of T_0 plants and analysed. Usually 20 plants were analysed, but more material might be needed in case of high copy numbers of the transgene(s).

Barley cv. 'Golden Promise' seeds were put on PRM medium without antibiotics, and germinated in a light chamber (24 °C, 16h photoperiod, 20 000 lux). Leaf material of small plantlets was taken and genomic DNA isolated.

Embryo rescue provides the possibility of reducing the time span needed to obtain the following generation and thus enables an earlier time point for their analysis. This way the time required for the ripening of seeds can be saved using immature embryos (dissected from surface sterilised seeds and put on CMR medium) to form small plantlets.

PCR, Southern Blot, *gus* histochemical assay and hygromycin leaf test were carried out the same way as explained by "2.2.5. Analysis of transgenic plants"

MATERIALS AND METHODS

2.5. Statistical treatment

The obtained data were analysed by parameter-independent Kruskal-Wallis One Way Analysis of Variance on Ranks (SigmaStat 3.0, SPSS Inc.,Chicago, IL, USA). Pairwise comparisons of the Agrobacterium/vector combinations against the control repetitions applied in parallel in the same experiment were performed. P values <0.05 were considered to indicate statistical significance.

Average absolute deviation values were calculated across the experimental repeats of each variant according to the formula given below, so as to intelligibly visualize the variation within treatments in the diagrams:

$$\frac{1}{n}\sum_{x_i=1}^{n} |x_i - \overline{x_q}|$$

where,

  n is the sample size

  x_q is the mean value

RESULTS

3. RESULTS

3.1. Binary vectors

Four types of binary vectors (p*hpt::gfp*, p*gus*, pTwin T and pTwin I) were used for the *Agrobacterium* based transformation experiments, containing either *hpt::gfp* (the selectable marker) and/or *gus* (the GOI), see Figures 2.1-2.3

The production of the binary vector containing the *gus* reporter gene (p*gus*) was carried out through integration of the *E. coli β-glucuronidase* gene, including the *StLS1* intron driven by the d35S promoter, into the *hpt*-free p6U binary vector.

The substrate for the production of the two Twin binary vectors (pTwin T and pTwin I), each harbouring two T-DNAs separated by left and right border sequences, was the modified pSB227 binary vector, in which two restriction sites were removed from its multiple cloning site (MCS). The d35S-*gus*-nos cassette was cut out from p*gus* vector and inserted into the modified MCS of the pSB227 plasmid. Two types of binary vector were obtained by this latter cloning step, where the orientation of the *gus* gene is different related to the *gfp* gene resulting tandem (pTwin T) and inverted (pTwin I) fragments.

Figure 3-1 presents derivatives of two *Agrobacterium tumefaciens* strains, AGL-1 and LBA4404, harbouring different binary vectors, which carry the gene-of-interest and/or the selectable marker gene.

RESULTS

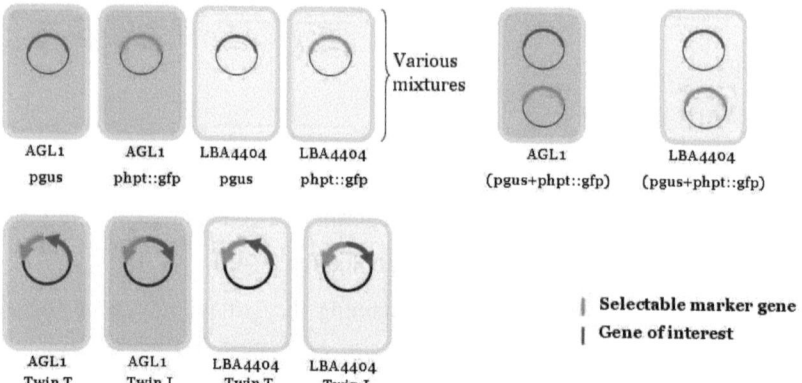

Figure 3-1
Four different types of binary vectors were transferred in two *Agrobacterium* strains resulting in a total of 14 *Agrobacterium*/vector combinations. For the sake of simplicity, the bacterial genetic background is not indicated.

3.2. Primary transgenic (T_0) plants

Barley genetic transformation was carried out through infection of immature embryos using two *Agrobacterium* strains (LBA4404 and AGL-1) according to the different experimental variants applied (Table 2-3). The embryo co-cultures comprised 14 different modes to employ the diverse *Agrobacterium* clones or mixtures. For these 14 different *Agrobacterium*/vector combinations (including methods with two plasmids/one strain, two plasmids in different strains, two plasmids in one *Agrobacterium* and Twin vectors harboring two T-DNAs) four types of binary vectors were used, in order to identify the most efficient combination(s) and mixture proportion(s) in terms of co-transformation and independent integration. The aim was to find the best variant(s) that will enable us to efficiently produce co-transformed primary T_0 plants with transgenes introduced in an unlinked manner, giving rise to selectable marker-free DH progeny which are instantly homozygous for the gene-of-interest (Figure 3-2).

First, regenerants were tested by PCR. Plant genomic DNA containing only the *hpt* produced a single 450 bp band, while co-transgenics resulted an additional 730 bp amplicon representing *gus*, the GOI (Figure 3-3)

RESULTS

A set of 606 regenerants carrying the selectable marker (coupled with the additional *gfp* reporter gene) was derived from 5,130 inoculated embryos (Appendix); these reflected the production of between one and 15 putative transgenic plants from each of 206 embryos (Diagram 3-1), Those plants which were able to grow under selective pressure conditions despite lacking the resistance gene are called escapes. Their proportion among the regenerants was 1.78%.

However, the 606 regenerants were produced by altogether 206 calli, because more than one plant was often obtained per callus, with numbers per callus being as high as fifteen. The sister plants derived from one single callus might either be genetically identical (clones) or represent independent lines. Initially only one plant per callus was considered independent, since molecular analysis was needed to determine the relationship among the progeny of the same callus.

Of the regenerants, 129 (derived from 50 embryos) also carried *gus* (Appendix). The stable integration of *hpt::gfp* and *gus* and their copy number was analysed by DNA gel blot, a procedure which was also able to recognize clonality among sister regenerants, in order to find out whether their analysis is worth to be conduced in future applications of the method.

RESULTS

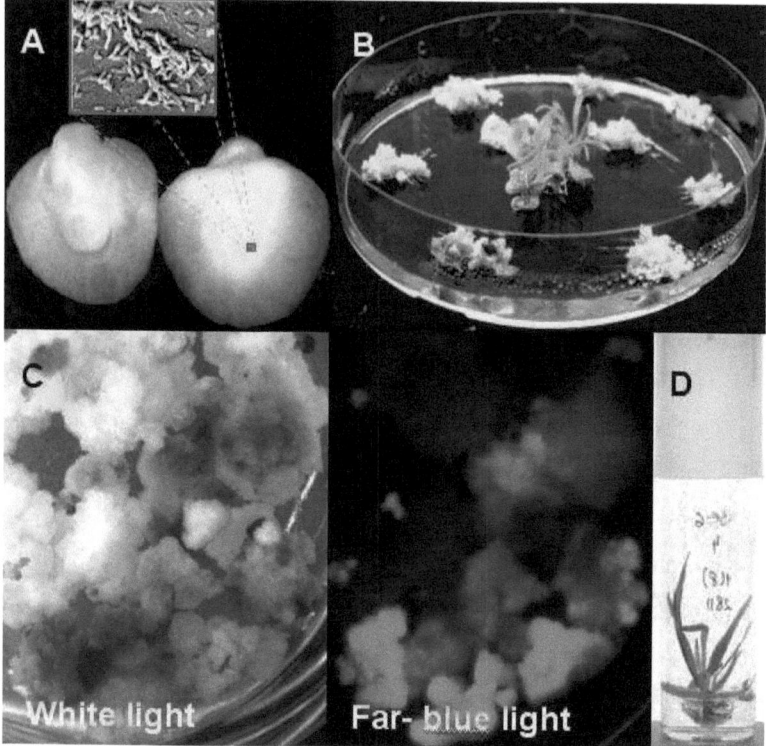

Figure 3-2
Co-transformation of immature embryos
  A. Immature barley cv. 'Golden Promise' embryos were used for *Agrobacteium*-mediated gene transfer
  B. Regenerating calli on selective medium (PRM+25 mg/l hygromycin B)
  C. Co-transformed calli showing expression of the reporter genes *gus* and *gfp*
  D. Primary regenerant (T_0)

RESULTS

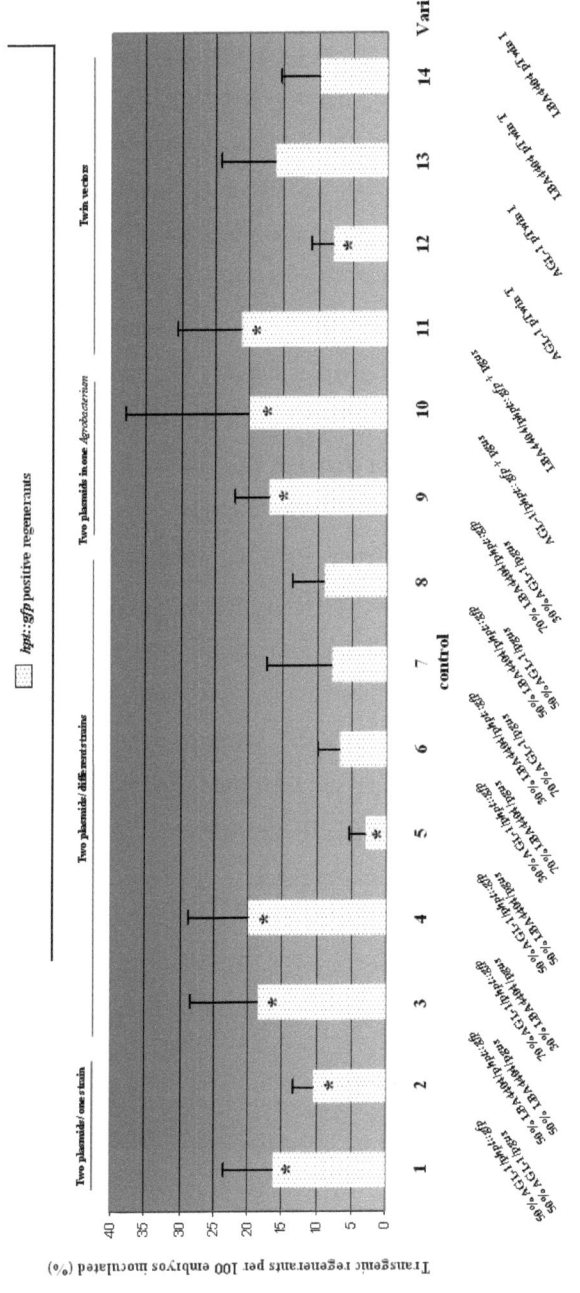

Diagram 3-1
Relative formation of hygromycin resistant regenerants.
Data represent the number of primary regenerants obtained per 100 inoculated barley 'Golden Promise' embryos, with their average absolute deviation values of the 3 repetitions per variant being indicated by error bars. Asterisks indicate a statistically significant difference (P<0.05) to those control repetitions that were conducted in direct comparison with the respective variant.

RESULTS

3.2.1. Evaluation of sister plants derived from the same embryo

As mentioned previously, occasionally several sister plants were produced per one callus, their clonal state can be determined by molecular analysis. In view of the PCR results co-transgenic sister plants were subject to Southern blot analysis, in order to identify, whether they represent genetically identical clones or independent multiple lines. Taking only independent lines into consideration was a prerequisite for an appropriate evaluation of co-transformation and segregation of the transgenes.

The regeneration pattern of the calli was followed through the experiments, in order to identify clones and independent lines derived from the same callus. Each of the calli which produced regenerants was given a number. The sister plants of a callus were numbered in alphabetical order, which means that plants indicated with the same number, but different letters stem from the same callus. Examples can be seen in Figure 3-3 and 3-4, e.g. 3a-d are four primary transgenic plants regenerated from callus number 3. As plant 3b is different from 3a, 3c and 3d it can already be deduced from the PCR test results that the regenerants produced by the same callus are not necesseraly clones. In principle the same applies to the two plants derived from callus number 4. By contrast, callus number 5 has produced three plants, all of which tested positive for both T-DNAs by PCR. These co-transgenic lines had to be further analysed by Southern blot to find out their individual integration pattern and transgene copy numbers in order to decide if they are to be considered clones or if there are independent transgenic lines among them.

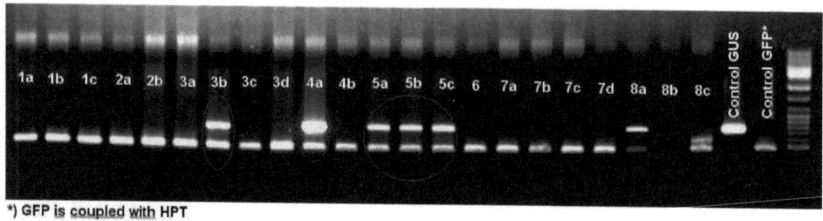

Figure 3-3
PCR analysis of the primary transgenic (T_0) plants. Numbers indicate different calli and letters the deriving regenerants. Co-transgenics referred to in the text are encircled in red.

RESULTS

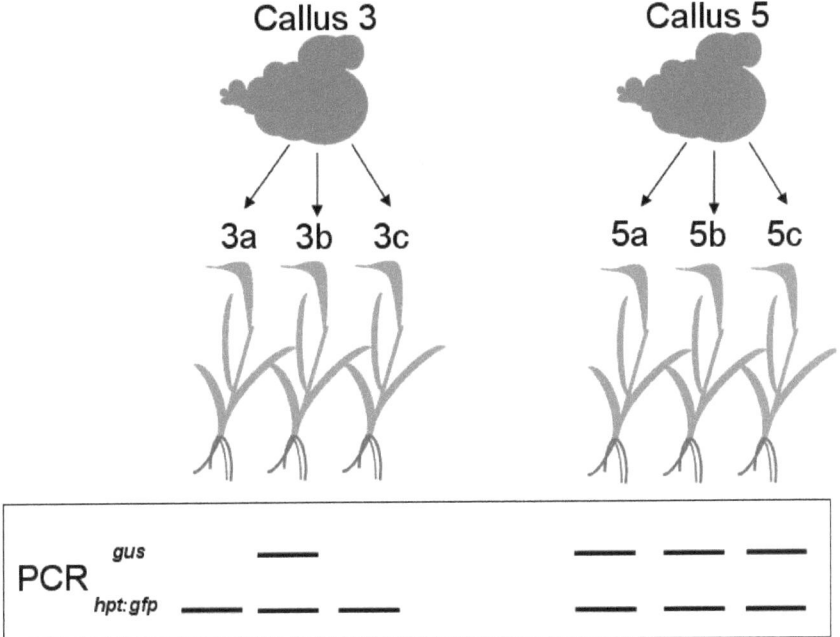

Figure 3-4
Comment on figure 3-3. PCR carried out on genomic DNA of plants obtained from co-transformation experiments might confirm genomic differences between regenerants coming from the same callus (callus 3). In other cases, the question of clonal state can only be answered with the use of other methods, such as Southern blot (callus 5).

The plantlets were put in glass tubes on regeneration medium containing hygromycin. In some cases it happened that in a tube containing one single plantlet, two or more other shootings appeared. The resultant plants were indicated by additional arabic numerals, e.g. 10a1 and 10a2, and were expected to be clones. However, it turned out from the PCR and Southern blot results that individual shoots can be escapes or might even stem from independent transformation or „supertransformation" (containing extra copy/copies) events (Figure 3-5).

RESULTS

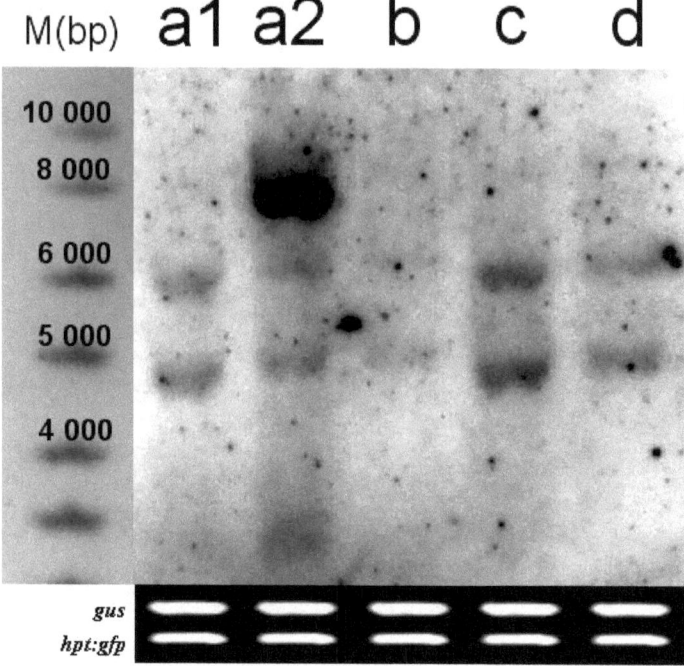

Figure 3-5
Southern blot revealing genetic differences among regenerants derived from the same callus. The bands represent copies of the selectable marker detected in five sister plants. Respective PCR results are shown below the Southern blot. It can be clearly seen that regenerant a1, b, c and d are clones, while plant a2, which grew together with a1 in one tube has at least two extra copies of the resistance marker. Note that the strong intensity of the uppermost band of plant a2 indicates a concatamer of multiple T-DNAs integrated.

Diagram 3-1 shows the distribution of the total number of primary transgenic regenerants among the 14 different variants, obtained per 100 inoculated embryos.

Co-transgenic sister plants regenerating from the same callus and later proven not to be clones by Southern blot were considered as additional independent transgenic lines in the calculation of the efficiencies, since these lines constituted additional candidates for the eventual obtention of marker-free lines. 30 % of the analysed co-transgenic sister families (6 out of 20) produced multiple independent lines (Table 3-1). About 30% of these families included non-identical transformants, so that in total, 55 independent co-transformed plants were obtained out of 228 independent transgenic lines.

RESULTS

Typically there were many not co-transgenic sister regenerants, which were not individually analysed for their integration pattern and transgene copy number, because they were no candidate progenitors for marker-free lines. As a consequence, it was impossible and not necessary to determine a total efficiency in independent transgenic plant generation. In those cases, where one callus produced more than one primary transgenic plants lacking the gene-of-interest (*hpt+, gus-*), the regenerants were considered as one single line. This means that the number of transgenic lines obtained from 100 inoculated embryos is underestimated by ca. 15%. Hereinafter, regenerants are considered as plants obtained from calli derived from *Agrobacterium*-mediated genetic transformation of immature embryos, without any further indication about their clonal state or relations. An independent line always refers to the sum of genetically identical clones or a single transgenic individual without any additional sister plants.

RESULTS

Table 3-1
Regeneration pattern of clones and independent lines of different variants.

Variant	No. of isolated IEs	Callus no.	No. of co-transgenic sister plants stemming from one callus	Any non-identical?	No. of lines proved independent per callus	No. of co-transgenic sister plants proved clones
2	270	1	5	yes	2	2+3
4	270	2	5	no	1	5
5	270	3	4	no	1	5
6	270	4	5	yes	3	1+2+2
7 control	1620	5	3	no	1	3
		6	3	no	1	3
		7	8	no	1	8
		8	2	no	1	2
		9	4	yes	3	1+1+2
9	270	10	5	yes	2	1+4
		11	5	no	1	5
		12	2	no	1	2
10	270	13	2	yes	2	1+1
		14	4	no	1	4
11	270	15	6	no	1	6
		16	11	no	1	11
		17	5	yes	2	1+4
12	270	18	2	no	1	2
13	270	19	5	no	1	5
14	270	20	15	no	1	15

RESULTS

3.2.2. Genetic transformation and co-transformation

There are various possibilities for presenting the effectiveness of the *Agrobacterium*-mediated gene transfer system to produce selectable marker-free transgenic barley lines. Three points of reference are mentioned hereinafter:

1. <u>Efficiencies</u> correlate with the *number of inoculated embryos*, e.g the transformation efficiency of a variant reveals the number of identified independent transgenic lines per hundred embryos.
2. <u>Frequencies</u> apply to the number of obtained *primary transgenic T_0 lines*, e.g. a co-transformation frequency represents the proportion of independent co-transgenic lines among the primary transgenic independent lines of the same variant.
3. <u>Rates</u> bear reference to *co-transgenic T_0 lines*, e.g doubled haploid production rate represents the proportion of co-transgenic T_0 lines producing doubled haploids in a variant.

Genetic transformation efficiencies were highly variable, ranging from 1.5 to 9.6% (Table 3-2, Diagram 3-2).

RESULTS

Table 3-2
Summary of results obtained from the 14 Agrobacterium/vector combinations tested.

Method	Agrobacterium strain/vector variant	A	Transformation efficiency (%)	B	Co-transformation efficiency (%)	C	D	GOI (+), SM-free production efficiency (%)
Two binary plasmids in two clones of the same Agrobacterium strain	1	11	4.1	2	0.7	1	0	0.0
	2	16	5.9*	5	1.9	3	0	0.0
	7 (replicates run with 1, 2)	5	1.9	3	1.1	3	1	0.4
Two plasmids in two different Agrobacterium strains	3	21	7.8	2	0.7	1	0	0.0
	4	26	9.6	2	0.7	2	2	0.7
	5	4	1.5	1	0.4	1	0	0.0
	7 (replicates run with 3, 4, 5)	15	5.6	4	1.5	4	1	0.4
Two plasmids in two different Agrobacterium strains	6	9	3.3	3	1.1	3	2	0.7
	7 (replicates run with 6, 8)	13	4.8	4	1.5	4	0	0.0
	8	13	4.8	1	0.4	1	1	0.4
Two plasmids in one Agrobacterium clone	9	24	8.9*	8	3.0	6	3	1.1
	10	15	5.6*	6	2.2	4	4	1.5*
	7 (replicates run with 9, 10)	5	1.9	2	0.7	2	0	0.0
Two T-DNAs in one binary vector in AGL-1	11	15	5.6*	6	2.2*	4	0	0.0
	12	6	2.2	1	0.4	1	0	0.0
	7 (replicates run with 11, 12)	2	0.7	0	0.0	0	0	0.0
Two T-DNAs in one binary vector in LBA4404	13	10	3.7	2	0.7	1	0	0.0
	14	6	2.2	2	0.7	1	0	0.0
	7 (replicates run with 13, 14)	12	4.4	1	0.4	1	0	0.0
	7 (sum of controls)	52	3.2	14	0.9	14	2	0.1

A: number of independent primary transgenic ($hpt::gfp$ positive) lines; B: Number of independent co-transgenic ($hpt::gfp$ and gus-positive) lines; C: number of independent co-transgenic lines producing green doubled haploid progeny; D: number of independent co-transgenic lines producing GOI-positive, selectable marker-free green doubled haploid progeny; *) higher efficiency of the *Agrobacterium*/vector combination on a statistically significant level ($P < 0.05$) as compared to the control repetition conducted in the same experimental run.

RESULTS

The results indicate that the most efficient variant from this aspect was variant 4 (1:1 mixture of AGL-1/p*hpt:gfp* and LBA4404/p*gus*). In this case 26 independent lines out of 270 inoculated immature embryos have been tested positive for the T-DNA of hygromycin resistance, which means a 9.6% transformation efficiency (Diagram 3-2). However, from the aspect of co-transformation efficiency this variant was not among the best (0.7 co-transgenic lines per 100 inoculated immature embryos).

A control variant was added to each embryo transformation experiment, because donor material quality cannot be equally provided over time. In the statistical analysis, the variants were compared only to those control cultures grown in the very same experiment. Considering all 14 variants in terms of transformation efficiency, statistically significant differences ($P<0.05$) between the control and variants 2, 5, 9, 10 and 11 were determined (Table 3-2, Diagram 3-2). However, because of the cumulative representation of all control data, the diagram does not necessarily show if a variant is significantly better or worse than its particular control. All of the variants, except for number 5, proved to be more efficient than their control in the pairwise comparison.

The highest co-transformation efficiency among all *hpt*-positive regenerants was achieved from variant 9 (two plasmids in one *Agrobacterium* clone method), for which eight out of the 270 explants (3.0%) gave rise to independent co-transgenic lines. Six lines carrying both *gus* and *hpt::gfp* were regenerated from variants 10 and 11 each, but a statistically significant difference with respect to the control could only be established for the latter (Table 3-2, Diagram 3-2).

Co-transformation frequency presents the proportion of co-transgenics related to the total number of obtained transgenic lines. Both T-DNAs were integrated in the plant genome in 40.0% of the lines in both variant 10 (two plasmids in LBA4404) and variant 11 (AGL-1 pTwin T). In both cases 6 lines out of 15 turned out to be co-transgenic for the selectable marker and the gene-of-interest (Diagram 3-3). On the whole the proportion ranged from 7.7% to 40.0%, but no statistical difference to the control was determined from the available data.

RESULTS

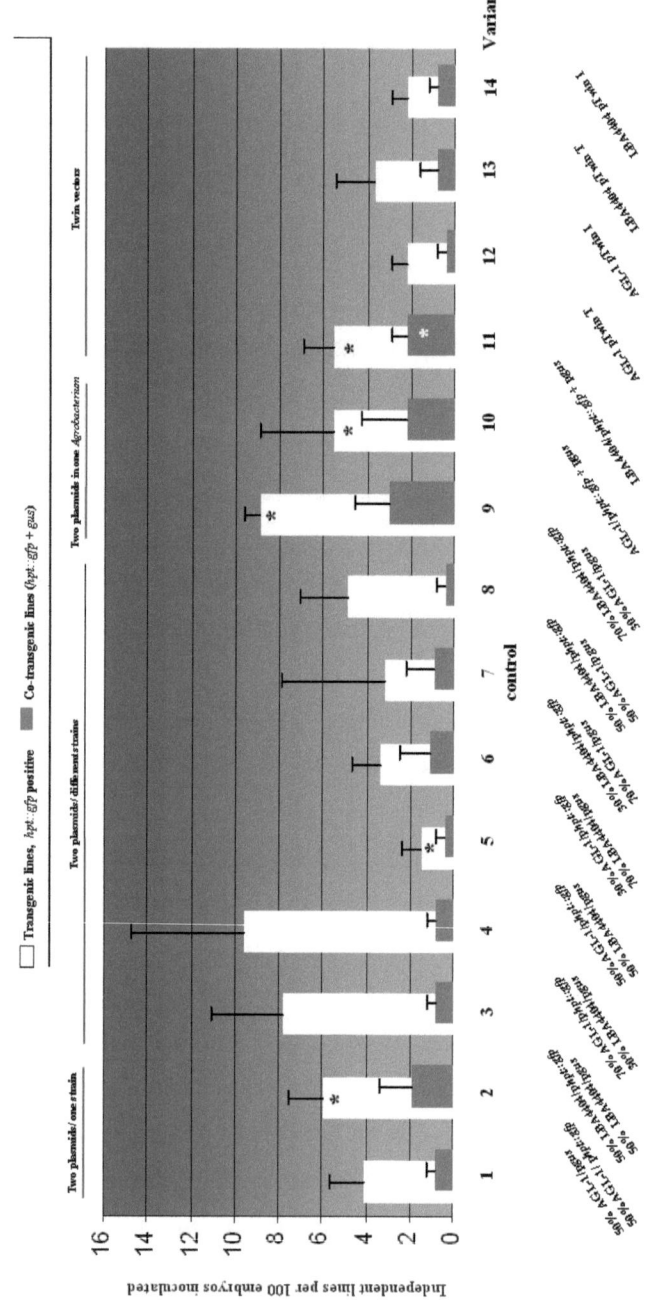

Diagram 3-2
Relative formation of hygromycin resistant co-transgenic lines.
Data represent the number of primary transgenic and co-transgenic independent lines obtained per 100 inoculated barley 'Golden Promise' embryos, with their average absolute deviation values of the 3 repetitions per variant being indicated by error bars. Asterisks indicate a statistically significant difference ($P<0.05$) to those control repetitions that were conducted in direct comparison with the respective variant.

RESULTS

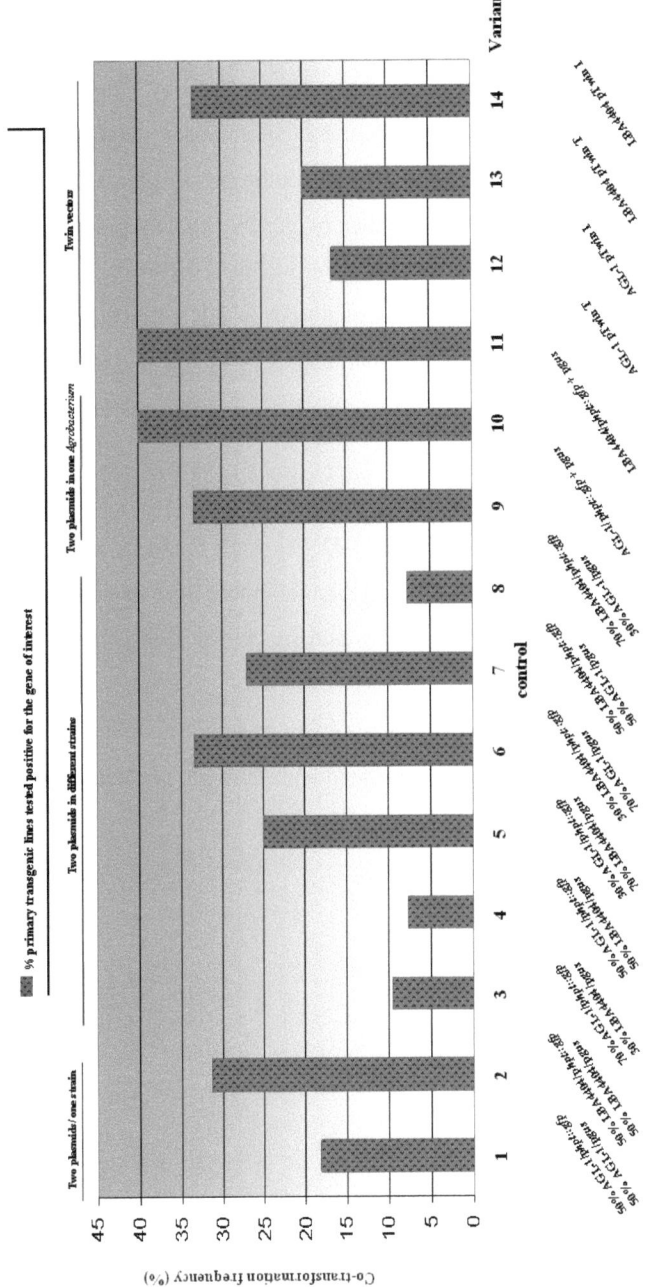

Diagram 3-3
Co-transformation frequencies:
The proportion of independent primary co-transgenic lines related to primary independent transgenic T_0 lines. No statistical significance was found between any of the treatments versus the parallel control variant 7.

RESULTS

3.2.3. Transgene copy numbers in primary co-transgenic lines

Stable integration of the selectable marker gene and the gene-of-interest in the plant genome was detected and their copy numbers were analysed by Southern blot in 91 primary co-transgenic regenerants. Table 3-3 summarises data on transgene integration in all analysed co-transgenic plants including sister plants. In view of the DNA hybridisation results, the final number of genetically independent lines was determined, depending on the clonal state of sister plants.

Of the 41 independent co-transgenic lines analysed by DNA gel blot the GOI *gus* was present as a single copy in 48.8%, as two copies in 17.1%, and as three or more copies in 34.1%. The equivalent frequencies for the selectable marker were 22.0%, 19.5% and 58.5% (Table 3-3, Diagram 3-4).

The different *Agrobacterium*/vector combinations did not markedly differ from each other. Lines carrying single copies of both T-DNAs were found in variants 2, 7, 11 and 12. On the other hand, occasionally very high numbers were detected, e.g. in one line of variant 7 more than ten copies of *gus* and six copies of the selectable marker were integrated in the plant genome.

However, using one enzyme for the detection of transgenic fragments bears the risk of underestimation of these data. Stronger band signals might represent two or more copies, which by chance have resulted in equal sizes.

RESULTS

Table 3-3
T-DNA copy numbers as detected by Southern blot in genomic DNA of co-transgenic T_0 plants including all sister lines. The plants identified as clones are considered together as a single independent line, which are represented by consecutive Roman numerals and placed in columns according to their copy numbers.

Variant	Proportion/ strain/ binary vector	No. of analysed (T_0) regenerants	Identified independent lines	T-DNA	No. of T-DNA copies determined by Southern blot (% of lines analysed)		
					1	2	>3
2	50% LBA4404/p*hpt::gfp* 50% LBA4404/p*gus*	6	3	*hpt::gfp*	I⁺ (33.3%)		II,III (66.7%)
				gus	I⁺,III (66.7%)	II (33.3%)	
3	70% AGL-1/p*hpt::gfp* 30% LBA4404pSB1p*gus*	1	1	*hpt::gfp*		I (100%)	
				gus	I (100%)		
4	50% AGL-1/p*hpt::gfp* 50% LBA4404/p*gus*	5	1	*hpt::gfp*			I (100%)
				gus			I (100%)
5	30% AGL-1/p*hpt::gfp* 70% LBA4404/p*gus*	4	1	*hpt::gfp*			I (100%)
				gus	I (100%)		
6	30% LBA4404/p*hpt::gfp* 70% AGL-1/p*gus*	5	3	*hpt::gfp*		II (33.3%)	I, III (66.7%)
				gus			I,II,III (100%)
7 control	50% LBA4404/p*hpt::gfp* 50% AGL-1/p*gus*	13	9	*hpt::gfp*	II,III VIII,IX (44.4%)	IV (11.2%)	I,V,VI VII (44.4%)
				gus	II,III,IV VI,VIII IX (66.7%)		I,V,VII (33.3%)
8	30% AGL-1/p*gus* 70% LBA4404/p*hpt::gfp*	1	1	*hpt::gfp*			I(T) (100%)
				gus	I(T) (100%)		
9	AGL-1/p*hpt::gfp* + p*gus*	16	8	*hpt::gfp*	II,VII(T) (25%)	I,IV,VIII VI(T) (37.5%)	III,V VI(T) (37.5%)
				gus	I,III,IV V (50%)	II,VII(T) (25%)	VI(T) VIII (25%)
10	LBA4404/p*hpt::gfp* + p*gus*	9	6	*hpt::gfp*		II,IV(T) (33.3%)	I⁺,III, V VI(T) (66.7%)
				gus		I⁺,II IV(T) (50%)	III,V VI(T) (50%)
11	AGL-1/pTwin T	23	5	*hpt::gfp*	I (20%)		II,III,IVV (80%)
				gus	I,II,III,V (80%)		IV (20%)
12	AGL-1/pTwin I	2	1	*hpt::gfp*	I (100%)		
				gus	I (100%)		
13	LBA4404/pTwin T	6	2	*hpt::gfp*			I,II (100%)
				gus		I (50%)	II (50%)

T) tetraploid line

RESULTS

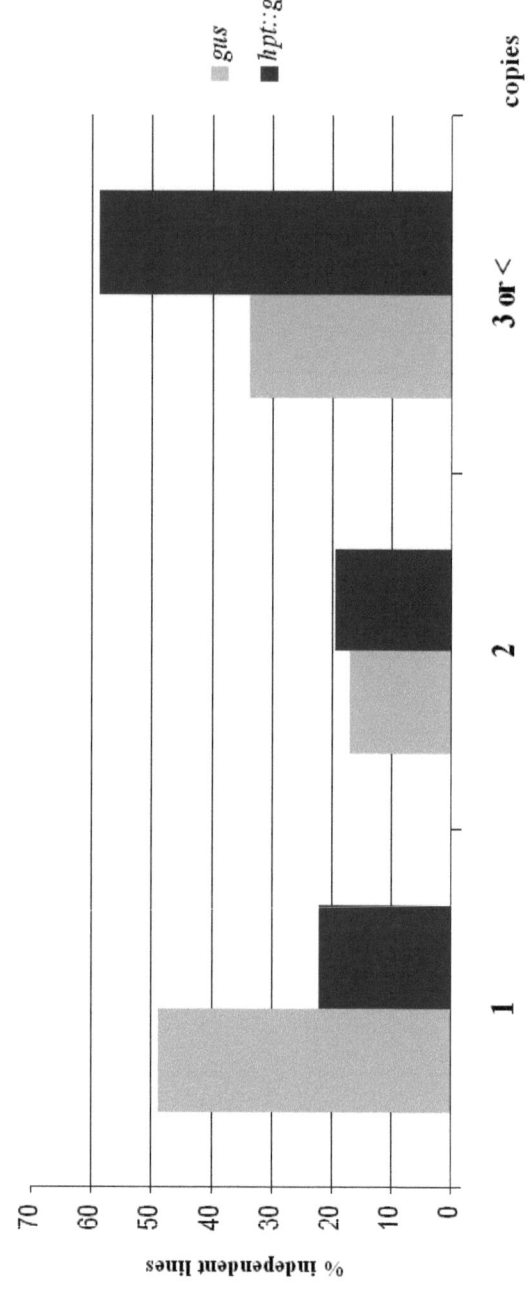

Diagram 3-4
Frequency of the two T-DNAs in the independent lines of the fourteen variants.

RESULTS

3.2.4. Ploidy variation among the primary transgenic regenerants

Agrobacterium-mediated gene transfer to immature barley embryos typically resulted in diploid pimary transgenic plants. However, occasionally some regenerants developed into plants which were abnormally tall, produced long, wide leaves and flowered late (Figure 3-6). They mature ca. one month later than their normal counterparts. The grain set in these plants is highly variable, though some produce good yield. Flow cytometry analysis revealed that such plants were tetraploid. Spontaneous genome doubling can occur when plants are regenerated from embryogenic cultures (Bregitzer et al. 1998; Choi et al. 2000; Gaponenko et al. 1988).

Microspore isolation from the spikes of tetraploid barley plants is possible and embryogenic pollen cultures can result in ample green regenerants. Progeny of such tetraploid T_0 individuals were dihaploid (2x, note the difference to doubled haploids) or, if spontaneous genome doubling happened, tetraploid (doubled dihaploids). Among all of the 14 variants producing a total of 606 primary transgenic regenerants, altogether 20 were tetraploid, which belonged to 5 independent lines. Among the pollen embryogenesis-derived T_1 plants of three independent tetraploid T_0-lines (ten plants) the *gus* gene segregated independently, producing transgenic selectable marker-free barley plants in diploid and tetraploid state (Table 3-4). Flow cytometry analysis also showed that the progeny regenerated from embryogenic pollen culture of one of the three T_0 lines included 18 diploid and two tetraploid individuals, while another produced eight diploids and six tetraploids.

Although those diploids are generated through haploid technology, further segregation of the GOI is still possible in the following generations, because the spontaneous genome doubling that has resulted in their 4x mother plants had been initiated from hemizygous somatic cells. Consequently, the 4x plants were also hemizygous and thus producing dihaploid pollen and respective pollen-derived dihaploid plants which expectedly segregate into 50% hemizygous, 25% homozygous transgenic and 25% azygous individuals.

RESULTS

Table 3-4
Distribution of tetraploid primary co-transgenic lines

Variant	Method	No. of tetraploid, independent co-transgenic lines	No. of the tetraploid lines producing SM-free transgenic progeny from pollen cultures
8	Two plasmids in two different *Agrobacterium* strains	1	1
9	Two plasmids in one *Agrobacterium*	2	1
10		2	1

The true number of tetraploid T_0 plants might be even higher than given here, because the occurrence of the phenomenon was not realised and followed from the beginning of the experiments and only co-transgenic regenerants were tested with regard to ploidy.

RESULTS

Figure 3-6
Comparison of barley plants of the same age, having diploid (on the left) and tetraploid (on the right) genomes. Tetraploid individuals tend to grow and mature slowly and produce larger grains.

3.3. Doubled haploids bred from co-transformant selections

Spikes of the potentially independent primary co-trangenic (T_0) lines were harvested, embryogenic pollen development was induced so as to generate doubled haploid (DH) populations, among which homozygous transgenic recombinants lacking the redundant selectable marker gene can be identified. In order to determine whether the model gene-of-interest segregated independently from the selectable marker in the T_1 generation, different analysis methods were carried out (PCR, hygromycin leaf assay and Southern blot).

RESULTS

3.3.1. Embryogenic pollen cultures

Embryogenic pollen cultures were induced from the immature spikes of 43 of the 55 co-transformants (Table 3-2) Clones facilitated the generation of DH lines from one independent line, since more material was available. The mean frequency of green doubled haploid regenerants from these cultures was 1.4 per spike (varying from 0.1 to 4.3), data not shown. This rate was sufficient to produce around 15 doubled haploid progeny per plant, as each produced an average of 10.7 harvestable spikes. (Figure 3-7).

RESULTS

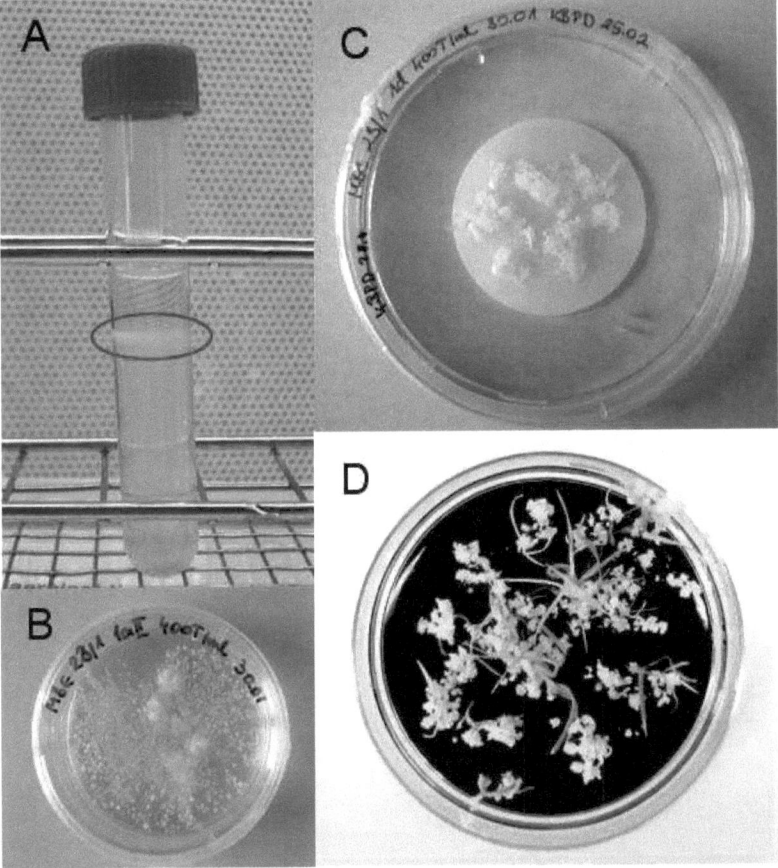

Figure 3-7
Isolation and pollen embryogenesis of barley microspores.
   A. Interphase containing the viable microspores
   B. Calli in KBP medium with immature wheat pistils
   C. Developing calli on solid KBP medium
   D. Regenerating calli producing haploid and doubled haploid plants

Diagram 3-5 presents green doubeld haploid production efficiencies among the 14 variants. Values range between 0.37 and 2.22%. The highest efficiency was performed by variant 9, because of the relatively high number of independent co-transgenic lines that were successful in regenerating plants from embryogenic pollen cultures. Although several variants produced a reasonal number of independent co-

RESULTS

transgenics giving rise to green DH plants, only variant 11 was significantly better as compared to its control.

On average, the proportion of green DH producing co-transgenic lines in relation to the number of independent co-transgenics of the same variant (doubled haploid production rate) ranged between 50 and 100% (Diagram 3-6). The variants do not significantly differ from their respective controls. In six variants, all of the primary co-transgenic plants gave rise to green doubled haploid progeny (4, 5, 6, 7, 8, 12).

RESULTS

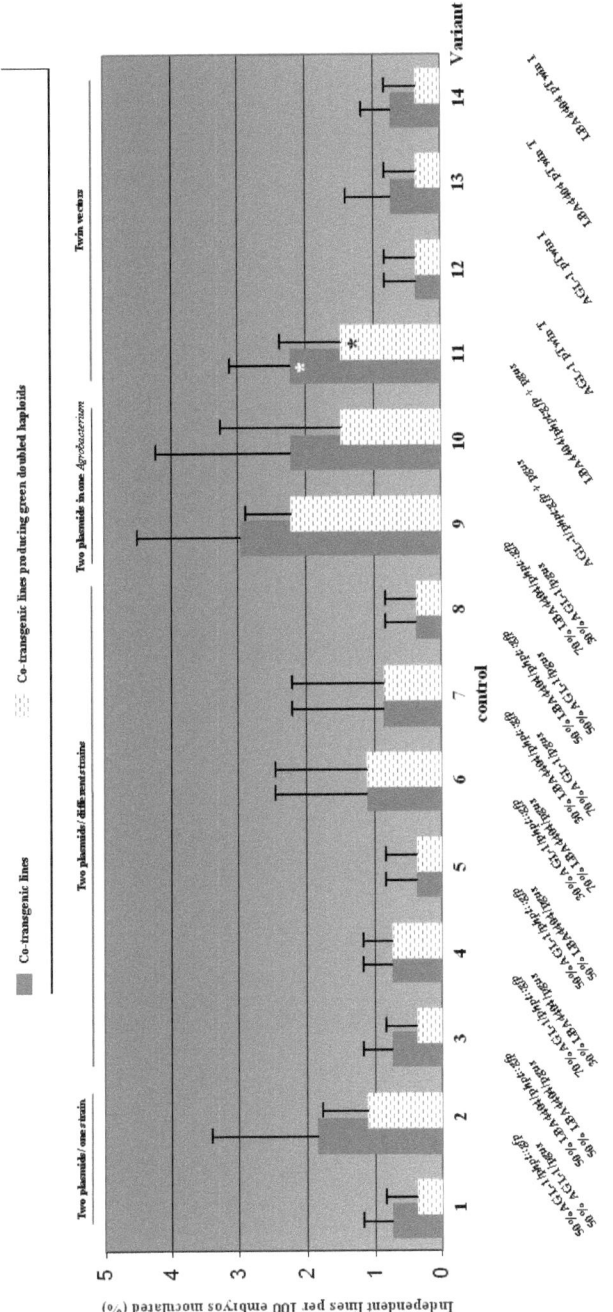

Diagram 3-5

Green doubled haploid production efficiency.

The columns represent the number of independent co-transgenic lines and their proportion producing green doubled haploid progeny per hundred inoculated embryos with the average absolute deviation across the 3 repetitions of a variant being shown by error bars. Asterisks indicate a statistically significant difference (P<0.05) to those control repetitions that were conducted in direct comparison with the respective variant, according to parameter-independent Kruskal-Wallis ANOVA on Ranks.

RESULTS

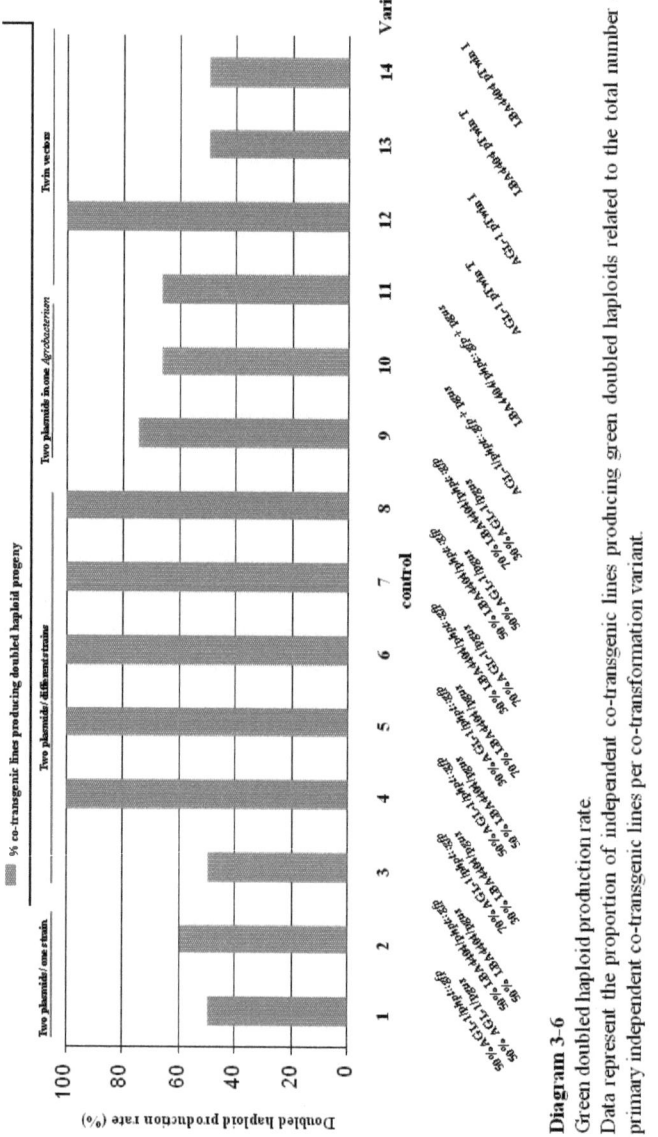

Diagram 3-6
Green doubled haploid production rate.
Data represent the proportion of independent co-transgenic lines producing green doubled haploids related to the total number of primary independent co-transgenic lines per co-transformation variant.

RESULTS

3.3.2. Environmental influence on the formation of doubled haploids

The outcome of immature barley embryo genetic transformation is highly dependent on environmental conditions. The obtained T_0 regenerants show phenotypic variability as well, e.g. in height, maturity, yield etc. A difference can also be seen when examining embryogenic pollen cultures even if they derive from the same barley genotype. Some cultures produce many microcalli, but in other cases hardly any cell division is observed. The number of green and albino regenerants is also highly variable.

It is difficult to determine which factor has greater influence on the DH production capacity of the embryogenic pollen cultures: seasonal variability or phenotypic deviations in the T_0 generation. Diagram 3-7 presents the average number of haploid and doubled haploid barley green and albino regenerants produced per spike during the four seasons of a year. Although the highest number of plants is produced in spring, the ANOVA on Ranks test shows no statistically significant difference between the median values among the four groups.

RESULTS

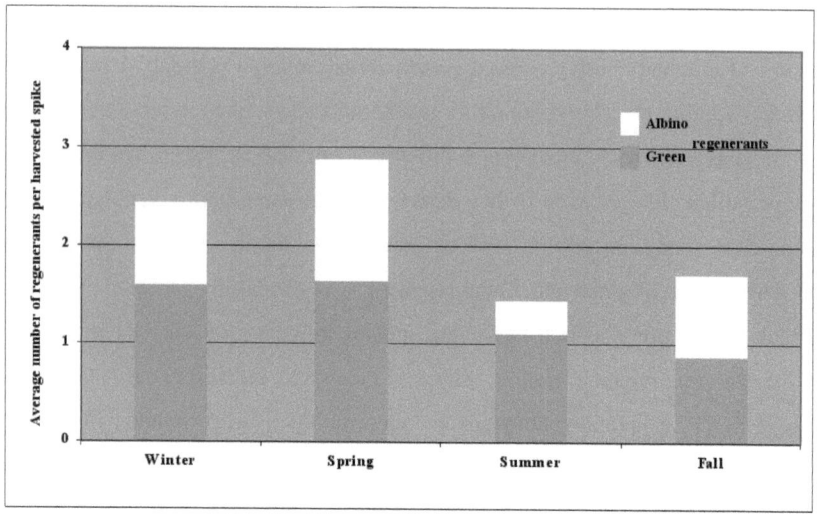

Diagram 3-7
Seasonal distribution of average numbers of green and albino regenerants per spike

3.3.3. Analysis of individual doubled haploid plants

Three assays (PCR, hygromycin leaf assay and DNA gel blot) were applied to determine whether *gus* segregated independently from the selectable marker among the doubled haploid progeny (Figure 1-1 C). First, those plants were selected by PCR which produced a single 730 bp *gus* band, but lacked the selectable marker, and *gus* staining was also carried out (Figure 3-8). Hygromycin leaf assay strengthened the PCR results, leaves of *hpt* sensitive plants bleached on medium containing a high level of hygromycin (Figure 3-9).

As mentioned before, a total of 55 co-transgenic independent lines were obtained. Only two of them proved to be sterile and produced neither doubled haploids nor seeds, data not shown.

The absence of linkage was detected in 31 of the 43 doubled haploid families. However, because some of the transformants carried at least one copy of both transgenes linked to

RESULTS

each other, the number of T_1 DH families in which *gus* could be separated from *hpt::gfp* was 14 (Table 3-2). Both T-DNAs (*gus* and *hpt::gfp*) segregated from each other in DHs derived from three independent co-transgenic lines.

Diagram 3-8 summarises the proportion of independent co-transgenic lines in the different variants producing embryogenic pollen culture-derived doubled haploids containing the gene-of-interest, but lacking the selectable marker.

From the aspect of efficient production of homozygous selectable marker-free transgenic barley cv. 'Golden Promise', the two plasmids in one *Agrobacterium* clone method proved to be the best, both variants showed statistically significant differences to their control variant.

In variant 9, three out of eight primary co-transgenic lines gave rise to SM-free and *gus* positive doubled haploid lines (equivalent to 1.1 lines per 100 embryos), while variant 10 (LBA4404/p*hpt::gfp* + p*gus*) produced an efficiency of 1.5 lines (4 of the 6 co-transgenic lines) per 100 inoculated embryos. Two plasmids in different strains was the only other method that generated GOI positive homozygous doubled haploids lacking the SM, namely variants 4, 6, 7 and 8.

No selectable marker-free transgenic individuals were identified among the DH families obtained from the two binary plasmids in two clones of the same *Agrobacterium* strain method (variants 1 and 2), variants 3 and 5 from the two plasmids in different strains method and the Twin variants (11, 12, 13, 14), see Table 3-2. Note, however, that it was possible to recover progeny from co-transformants which carried *hpt::gfp* but not *gus*, showing that the two T-DNAs can be separately inserted even from a Twin vector (data not shown). In these transformants, it appears that multiple (three or more) copies of *hpt::gfp* had been inserted, with one or more of these insertion sites also containing a copy of *gus*. Three or more copies of the *hpt::gfp* were also found frequently in such lines.

RESULTS

Selectable marker-free transgenic doubled haploid production rate presents the proportion of co-transgenic T_0 lines producing hygromycin sensitive, GOI positive regenerants per number of independent co-transgenics (Diagram 3-9). Values ranged between 14.3% and 100%.

RESULTS

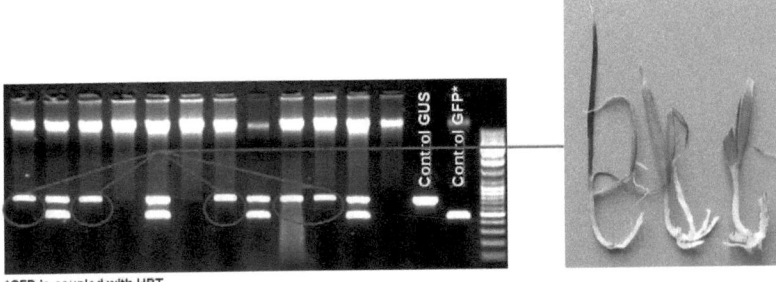

*GFP is coupled with HPT

Figure 3-8
Identification of selectable marker-free transgenic progeny

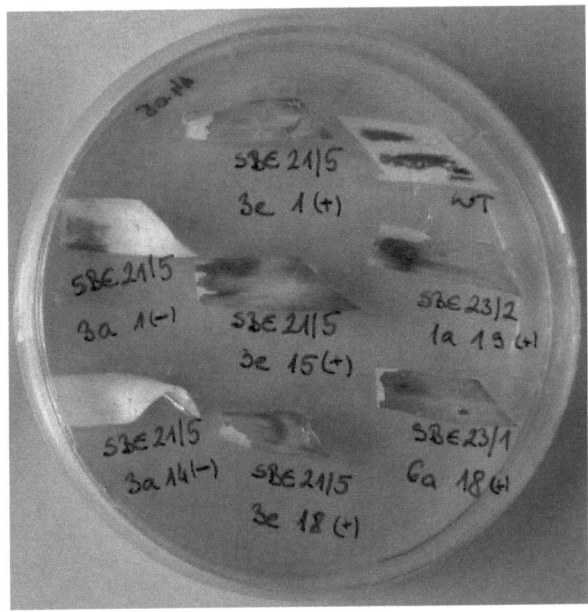

Figure 3-9
Leaf assay showing presence or absence of hygromycin resistance gene expression (-) PCR *hpt* negative, (+) PCR *hpt* positive, WT- wildtype

RESULTS

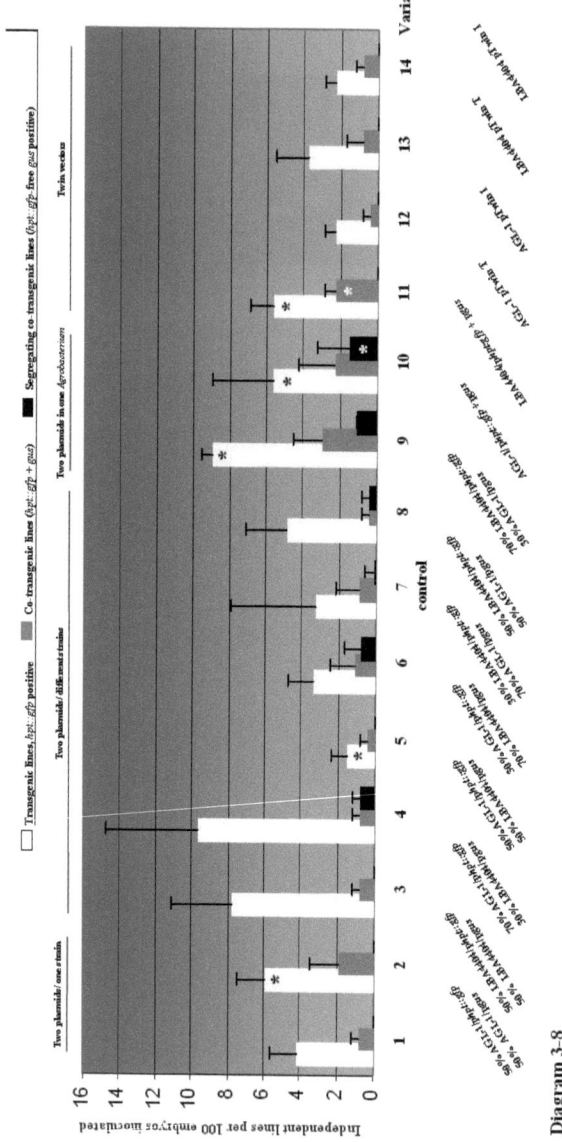

Diagram 3-8
Formation of independent primary transgenic, co-transgenic and selectable marker-free transgenic lines.
Data represent the number of independent primary transgenic and co-transgenic lines as well as those lines producing independent selectable marker-free GOI positive green doubled haploid progeny as quantified per hundred inoculated embryos, with their average absolute deviation being shown by error bars. Asterisks indicate a statistically significant difference (P<0.05) to the respective control repetitions according to the parameter-independent Kruskal-Wallis ANOVA on Ranks.

RESULTS

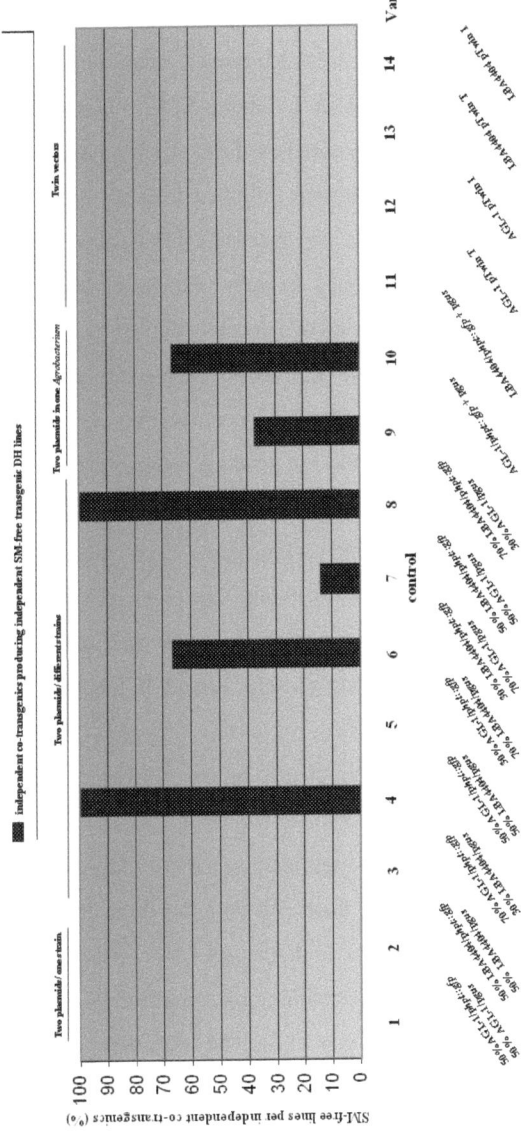

Diagram 3-9

Selectable marker-free transgenic doubled haploid production rate, showing the proportion of independent primary co-transgenic lines generating *hpt::gfp* negative *gus* positive progeny per total number of co-transgenic T_0 lines

RESULTS

3.3.4. T-DNA segregation in populations of doubled haploids

Southern blot profiles were informative with respect to transgene copy numbers in primary T_0 transgenic plants (Table 3-3). Moreover, analysis of the T_1 generation reveals the relation of the integrated DNA fragments in the plant genome, i.e. whether they are coupled in linkage groups, or located on different chromosomes.

Usually only one of the two transgenes was unlinked from all copies of the other and produced DH regenerants carrying only one type of T-DNA, as was the case in ca. the half of the independent co-transgenic lines. In such case, there are often multiple copies of the two T-DNA types integrated in linkage groups coupled to each other. If just one of the copies is unlinked from all others, segregants exclusively carrying the respective T-DNA type are obtained, but no plants tested positive only for the other type. This means, that a given co-transgenic line did not produce progeny, among which both only GOI and only SM positive genotypes were found.

However, in 43.6% of the independent co-transgenic lines, the transgenic fragments were integrated in linkage groups, without any uncoupled one, during the *Agrobacterium*-mediated gene transfer. In such cases the chance for separation decreases sharply. The likelihood that closely linked T-DNAs segregate through recombination during meiosis is very low.

Linked integration of the transgenes can clearly be followed by Southern blot analysis of T_1 populations obtained either from doubled haploid or sexual progeny (Figure 3-10). Table 3-5 and Diagram 3-10 present the segregation pattern of the selectable marker and the model gene-of-interest in several lines. Among the three independent DH lines, where segregation of both types (*gus* and *hpt::gfp*) of T-DNA occurred, one line contained a single copy of both the selectable marker and the GOI in their genome. In one line one copy of the GOI and 2 copies of the SM were detected, but the latter were coupled to each other as it turned out from the Southern blot results, and one co-transgenic plant had two copies of the GOI and one *hpt::gfp*.

RESULTS

One of the transformants from variant 8 produced exclusively selectable marker-free *gus* positive progeny, even though it was known that the transformation event involved five copies of the *hpt::gfp* transgene.

In the Twin variants only segregation of the *hpt::gfp* T-DNA was observed, and mostly the detected number of copies was high.

Interestingly, even if there are many copies detected by Southern blot, most of them were integrated linked to each other. These linkage groups will likely segregate as one locus in the T_1 generation. According to DNA gel blot analysis of DH progeny, one derivative of variant 13, LBA4404/pTwin T, included a multi-event involving the integration of eight *hpt::gfp* T-DNA copies at five separate loci. Hygromycin resistant homozygous DH progeny lacking the GOI were obtained, but no marker-free *gus* positive ones were found in the doubled haploid T_1 population. Probably the two *gus* T-DNA copies were integrated in a linked manner to at least one of the selectable marker gene copies.

Due to the frequent linkage of both types of T-DNA, the vast majority of DH lines were either co-transgenic for both T-DNAs or lacking any transgenic fragments.

RESULTS

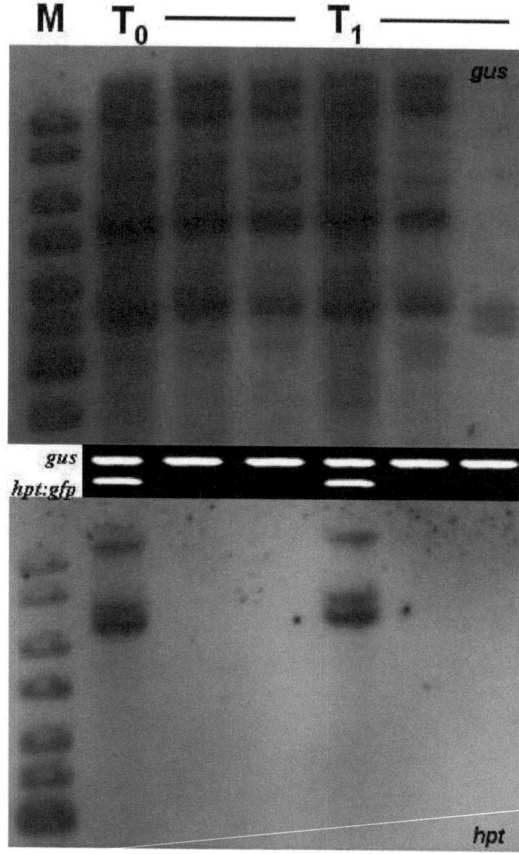

Figure 3-10
Example for the segregation pattern of both types of T-DNAs among doubled haploid T_1 progeny of a co-transgenic T_0 plant using PCR and Southern blot. *Gus* T-DNAs were integrated in two linkage group loci, one of which proved not to be coupled with the one *hpt* linkage group.

RESULTS

Table 3-5
Segregation pattern of the two T-DNAs in the T_1 generation

Variant	Proportion/ strain/ binary vector	Identified independent lines	T-DNA	No. of T-DNA copies determined by Southern blot		
				1	2	>3
2	50% LBA4404/p*hpt::gfp* 50% LBA4404/p*gus*	3	*hpt::gfp*	I*⁺		II*(3L+1S) III*(4L+1S)
			gus	I*⁺,III	II	
3	70% AGL-1/p*hpt::gfp* 30% LBA4404/p*gus*	1	*hpt::gfp*		I*(2L)	
			gus	I		
4	50% AGL-1/p*hpt::gfp* 50% LBA4404/p*gus*	1	*hpt::gfp*			I(3L)
			gus			I* (4L+1S)
5	30% AGL-1/p*hpt::gfp* 70% LBA4404/p*gus*	1	*hpt::gfp*			I*(2L+1S)
			gus	I		
6	30% LBA4404/p*hpt::gfp* 70% AGL-1/p*gus*	3	*hpt::gfp*		II(2L)	I(2L+1S),III
			gus			I*(>3S) II*(>3S) III(5L)
7 control	50% LBA4404/p*hpt::gfp* 50% AGL-1/p*gus*	9	*hpt::gfp*	II*,III VIII,IX	IV	I,V,VI*VII
			gus	II*,III*, IV,VI VIII,IX		I,V VII*
8	70% LBA4404/p*hpt::gfp* 30% AGL-1/p*gus*	1	*hpt::gfp*			I
			gus	I*		
9	AGL-1/p*hpt::gfp* + p*gus*	8	*hpt::gfp*	II*,VII	I*(2L),IV* VIII	III*,V*,VI*
			gus	I*,III,IV V	II*,VII*	VI,VIII
10	LBA4404/p*hpt::gfp* + p*gus*	6	*hpt::gfp*		II(2L) IV(2L)	I⁺,III(2L+1S)V (4L),VI
			gus		I⁺,II* IV*(2S)	III* V*(3L+1S) VI
11	AGL-1/pTwin T	5	*hpt::gfp*	I		II*,III*, IV,V*
			gus	I,II,III,V		IV
12	AGL-1/pTwin I	1	*hpt::gfp*	I		
			gus	I		
13	LBA4404/pTwin T	2	*hpt::gfp*			I*(3L+2L+3S) II
			gus	I		II

*) Plants that segregated in the T_1 for the respective T-DNA are indicated by an asterisk
+) indicate if no doubled haploids were obtained from a line, and sexual T_1 individuals were analysed
T-DNAs introduced in a linked manner in one locus are indicated by L, with the number of copies in the linkage group, e.g 2L means two copies of the transgene in a linkage group. Additional single copies are indicated by S.

RESULTS

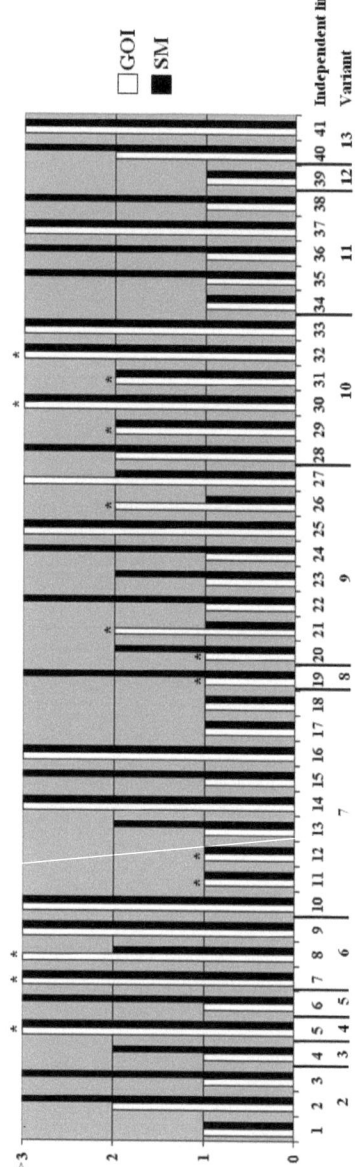

Diagram 3-10
T-DNA copy number in independent co-transformants, and their segregation in the DH T1 generation as determined by DNA gel blot analysis.
*) T1 families which included segregants carrying the GOI but no SM

RESULTS

3.4. Sexually generated T1 lines

No doubled haploid regenerants were obtained from the embryogenic pollen cultures of 12 primary co-transgenic independent lines, belonging to variants 1 (one line), 2 (two lines), 3 (one line), 9 (two lines), 10 (two lines), 11 (one line), 12 (one line), 13 (two lines). Among them only one line (variant 2, 50% LBA4404/p*gus* and 50% LBA4404/p*hpt:gfp*) was identified, where the resistance gene and the gene-of-interest were integrated in an unlinked manner, giving rise to transgenic selectable marker-free sexual T_1 plants which need to be further analysed in T_2 and probably T_3 to generate and identify a respective homozygous line (Table 3-5, Figure 3-11).

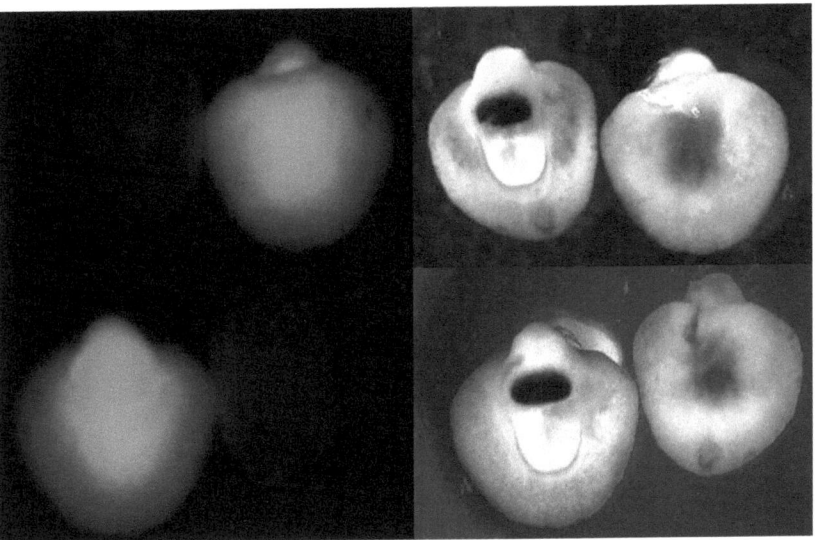

Figure 3-11
Embryos of a co-transgenic plant showing independent segregation of the *hpt::gfp* T-DNA and the *gus* T-DNA.

RESULTS

3.5. Time frame of the established method

Table 3-6 presents a time line for the production of doubled haploid transgenic barley selections, achieved using *Agrobacterium*-mediated co-transformation followed by immature pollen culture-based generation of doubled haploid progeny. The whole process took about 43 weeks to move from the dissection of immature embryos to the identification of homozygous transgenic, selectable marker-free plants.

RESULTS

Table 3-6
Time plan for the production of doubled haploid transgenic barley cv. 'Golden Promise' from embryogenic pollen culture of co-transgenic T_0 plants generated by *Agrobacterium*-mediated gene transfer to immature embryos

Steps in succession	Time interval
***Agrobacterium*-mediated transformation of immature embryos**	
Co-culture of immature embryos with agrobacteria	60 hours
Callus induction	2x2 weeks
Plant regeneration	4x2 weeks
Growth and maturation of the primary co-transgenic T_0 plants	ca. 17 weeks
Production of doubled haploid T_1	
Spike harvest from co-transgenic plants and cold treatment	3 weeks
Starvation treatment of isolated immature pollen	2 days
Callus formation in liquid KBP medium	3 weeks
Callus development on solid KBPD medium	2 weeks
Plant regeneration and identification of selectable marker-free transgenic individuals	3x2 weeks

4. DISCUSSION

4.1. Efficiency of the established method

Monocotyledonous plants do not naturally belong to the hosts of *Agrobacterium tumefaciens*, but efficient protocols were developed for genetic transformation of such non-hosts, with barley among them (Fang et al. 2002; Hensel et al. 2008; Kumlehn et al. 2006; Murray et al. 2004; Patel et al. 2000; Stahl et al. 2002; Tingay et al. 1997; Travella et al. 2005; Trifonova et al. 2001; Wang et al. 2001; Wu et al. 1998). The pioneer protocols used *bar* and *hpt* as selectable marker. In the present study we used *hpt*, according to the protocol of Hensel and Kumlehn (2004), because due to its effectiveness in contrast to other selection systems the proportion of non-transgenic escapes (1.78% in this study) can be significantly reduced.

Successful co-transformation requires a very efficient genetic transformation protocol and the induction of sufficient independent events to allow for the separation of the selectable marker from the GOI via conventional segregation. This thesis presents experiments in which a range of co-transformation strategies were compared, involving methods using mixtures of *Agrobacterium tumefaciens* strains (Coronado et al. 2005; De Block and Debrouwer 1991; Komari et al. 1996; McKnight et al. 1987), two plasmids in one *Agrobacterium* clone (Daley et al. 1998; De Framond et al. 1986; Komari et al. 1996) and a vector, Twin, harboring two T-DNAs (Komari et al. 1996; Matthews et al. 2001; Stahl et al. 2002). The latter method seemed very promising for the purpose of establishing the method presented in this thesis. Komari et al. (1996) co-transformed rice and tomato plants using a super binary vector, where the two T-DNA regions were separated by a large, at least 15 kb section, on a huge 50-55 kb plasmid obtained by homologous recombination. The advantage of this system is that it gives the possibility for the substitution of various GOIs, opposed to the conventional vector systems. On the other hand, the vector is very large and cumbersome to work with. Co-

DISCUSSION

transformation frequency was between 47-85%, more than half of the lines were SM-free *gus* positive. They also conducted experiments with *Agrobacterium tumefaciens* strain mixtures, but co-transformation efficiency was lower.

Two adjacent T-DNAs, separated by left and right border regions, integrated in a standard binary vector were transformed by Matthews et al. (2001) into barley. The transgenes were divided only by a small plasmid region (850 bp). Transgenic lines were produced with 2-12% transformation efficiency, 66% co-transformation frequency, 24% of the co-transgenics produced selectable marker-free progeny containing the GOI, resulting in 16% useful independent co-insertions events.

Table 4-1 summarises previous techniques for the generation of co-transformants, in which the GOI and SM gene can be segregated in the successive generation. Co-transformation frequency values and the proportion of selectable marker-free lines with regard to the number of co-transgenics showed great variability among the reports, but it must be mentioned that different species were subject to the co-transformation experiments. It has been widely discussed (Depicker et al. 1985; Komari et al. 1996; Matthews et al. 2001) that using single strain methods (two T-DNAs in one *Agrobacterium* clone, Twin vectors), the co-transformation efficiency is much higher than applying mixture methods (two plasmids/one strain, two plansmids in different *Agrobacterium* strains). However, this hypothesis was contradicted by McKnight et al (1987).

DISCUSSION

Table 4-1
Delivery of T-DNAs to plants using *Agrobacterium*-mediated co-transformation for the generation of selectable marker-free transgenic lines.
n.d. not determined

Species	Proportion of co-transgenic lines per all transformants (%) Co-transformation frequency	Selectable marker-free lines per co-transgenics (%) SM-free transgenic production rate	Method	*Agrobacterium* strain	GOI	SM	Reference
Nicotiana tabacum	49	19	Mixture method: Two plasmids/one strain (1:1)	*A. rhizogenes* A4	*nos*	*npt*	McKnight et al. 1987
Brassica napus	39-85	n.d. linked integrations	Mixture method: Two plasmids/one strain (1:1)	*A. tumefaciens* C58	*bar*	*npt*	Block and DeBrouwer 1991
Brassica napus	62	40	Single strain method: Two plasmids in one *Agrobacterium* clone	*A. tumefaciens* LBA4404	*gus*	*npt*	Daley et al. 1998
Nicotiana tabacum and *Oryza sativa*	47-85	50	Single strain method: Two plasmids in one *Agrobacterium* clone	*A. tumefaciens* LBA4404	*gus*	*hpt npt*	Komari et al. 1996
	0-35	ca. 50	Mixture method: Two plasmids/one strain (1:1 and 3:1)	*A. tumefaciens* LBA4404	*gus*	*hpt npt*	
Hordeum vulgare cv. 'Golden Promise'	66	24	Twin vector	*A. tumefaciens* AGL-1 and AGL-0	α-amylase α-glucosidase	*hpt*	Matthews et al. 2001
Hordeum vulgare cv. 'Golden Promise'	34.6	5.6	Mixture method: Two plasmids/two strains	*A. tumefaciens* AGL-1 and LBA4404	*gus*	*hpt*	Coronado et al. 2005
Hordeum vulgare cv. 'Golden Promise'	7.7-40	67 (Variant 10)	Mixture methods Single strain methods	*A. tumefaciens* AGL-1 and LBA4404	*gus*	*hpt*	Present study

94

DISCUSSION

Barley cv 'Golden Promise' genetic transformation efficiencies presented in this study ranged from 0.7 to 9.6% among the different variants (Table 3-2, Diagram 3-2). These rates are comparable with current protocols based on immature embryo explants of 'Golden Promise' (Goedeke et al. 2007; Hensel and Kumlehn 2009). Transgenics carrying either the selectable marker only or both transgenes were recovered from each strain/vector variant. It was not possible to carry out the immature barley embryo genetic transformation of all the variants at once, division of the experiments was necessary. Due to this fact, a control variant (1:1 mixture of AGL-1/p*gus* and LBA4404/p*hpt::gfp*) was applied, which makes transformation and co-transformation efficiencies, frequencies and rates comparable to a standard, and suitable variants can be selected.

Two percent of the inoculated embryos gave rise to more than one regenerant. A detailed analysis of the co-transgenic sister plants regenerating from the same callus revealed that they were mostly clones. According to the obtained data summarized in Table 3-1, the probability that two regenerants randomly chosen from such multiple plant producing calli belong to different independent lines is 9.3%. Due to the tremendous additional effort necessarily required for their analysis, it is recommended to discard those lines in future applications.

The overall transformation efficiency was not correlated with the co-transformation efficiency, and only variant 11 (AGL-1/pTwin T) differed significantly from its control combination for both these aspects (Table 3-2, Diagram 3-2). Variants 3 and 4 of the two plasmids in different strains method were effective with respect to the formation of independent transgenic T_0 lines per 100 immature embryos. In the same time, a high number of *hpt* positive lines were obtained from the parallel control variant, which suggests that the quality of the donor material was better than usual, thus the two variants did not prove to be outstanding in transformation efficiency. However, their co-transformation efficiency values were not among the best. The lack of association

DISCUSSION

between genetic transformation and co-transformation of barley cv. 'Golden Promise' is probably due the fact, that presence of the second T-DNA, the gene-of-interest, does not provide the plants with any benefit in the regeneration process, therefore there is no selection for it.

Another way to test the effectiveness of the system is to show, what proportion of the lines were co-transgenic in comparison to the total number of obtained antibiotic resistant lines. This is expressed by the co-transformation frequency (Diagram 3-3). Variants 10 (two plasmids in LBA4404) and 11 (AGL-1/pTwin T) reached the highest value (40%). This is lower than the 66% co-transformation frequency result published by Matthews et al. (2001). Although the mean values showed a range from 7.7% up to 40.0%, the high average absolute deviation values indicate that the data is not necessarily conclusive. Further repeats of the experiments would be needed to increase the statistical power of the calculations.

In the different variants, all together 55 primary co-transgenic independent lines were identified, 72.7% of these selected plants gave rise to successful embryogenic pollen cultures.

DH production efficiency was expectedly unaffected by the co-transformation method, and thus the most DH populations were obtained from variants with a high output of co-transgenics thus the highest green doubled haploid production efficiency, belonged to variant 9 (Diagram 3-5).

Doubled haploid production rates, presenting the proportion of primary co-transgenic lines successfully producing green DH progeny, ranged between 50% and 100% (Diagram 3-6). However, the proportion of independent lines producing green DHs per 100 isolated immature embryos must be taken into consideration as well. Only one primary co-transgenic plant was obtained in each of three variants (5, 8, and 12), all of which produced green doubled haploids (100% green DH production rate), while 270

DISCUSSION

embryos have to be inoculated to obtain only one of such co-transgenic line, and thus 0.4% of the embryos produced green DHs.

T-DNA segregation events were observed in more than half of the primary co-transgenic T_0 plants. However, only every second of these segregating lines produced *gus* positive plants lacking the selectable marker, because frequently the two types of T-DNA were integrated in linkage groups with at least one extra copy of the selectable marker unlinked from all others. In such case, the segregants carrying exclusively the hygromycin resistance gene are obtained, but no *gus* positive, SM-free plants were found. All but one of the GOI segregating T_0 regenerants generated homozygous DH populations of embryogenic pollen culture origin. From a single line selectable marker-free *gus*-positive T_1 segregants were exclusively obtained.

Great variability is shown among the different variants of the applied methods for the number of SM-free transgenic lines per 100 embryos inoculated (Diagram 3-8). Only six of the 14 *Agrobacterium* strain/vector combinations gave rise to the desired class of progeny, and neither the 'two plasmids present in two clones of the same strain' method nor the Twin method produced any selectable marker-free doubled haploids containing the GOI. The most likely reason for the lack of such plants in these variants is the comparatively high number of integrated selectable marker T-DNAs, which increases the probability that one or several of them are linked to the gene-of-interest.

The single method that yielded outstanding numbers of SM-free transgenic DH lines was the two plasmids in one *Agrobacterium* clone (single strain methods), irrespective of the identity of the *Agrobacterium* strain. However, the recovery efficiency of selectable marker-free *gus* positive doubled haploid progeny per 100 inoculated explants was significantly greater than that achieved in the control variant only for LBA4404 (Table 3-2, Diagram 3-8). Using this variant 10, selectable marker-free *gus* positive doubled haploids were produced from 1.5 out of 100 explants. The results of

DISCUSSION

the two plasmids in different strains method were diverse, up to 0.7 selectable marker-free transgenic lines per 100 isolated immature barley embryos (variant 4 and variant 6). Selectable marker free transgenic DH production rates, the proportion of co-transgenic lines giving rise to hygromycin sensitive and *gus* positive plants, were 100% in variants 4 and 8. However, a total of only one co-transgenic line had been obtained by each of theses variants (Diagram 3-9).

Variant LBA4404/p*hpt::gfp* + p*gus*) is more favourable from the aspect of selectable marker free transgenic DH production rate, in comparison to AGL-1/p*hpt::gfp* + p*gus*, because the former variant had 66.7% of the primary co-transgenic plants (4 out of 6) producing DH progeny lacking the selectable marker, but tested positive for the model GOI, while in the latter this ratio was only 37.5% (3 out of 8).

The advantage of the here presented *Agrobacterium*-mediated gene transfer system is that it provides a comparatively efficient method to generate selectable marker-free plants which are instantly homozygous for the gene-of-interest. They are produced as early as in the T_1 generation, which is a faster way when compared to the conventional selection of homozygous sexual plants in the T_2 generation, and the whole generation process can be conducted within eleven months.

This can be demonstrated through a comparison for the most efficient method 'two plasmids in one *Agrobacterium* clone' (variants 9 and 10) between the recovery of selectable marker-free, *gus* positive individuals using haploid technology and the expected outcome of the same transgenic situation based on selection for homozygosity in the sexual T_2 generation is given in Diagram 4-1.

DISCUSSION

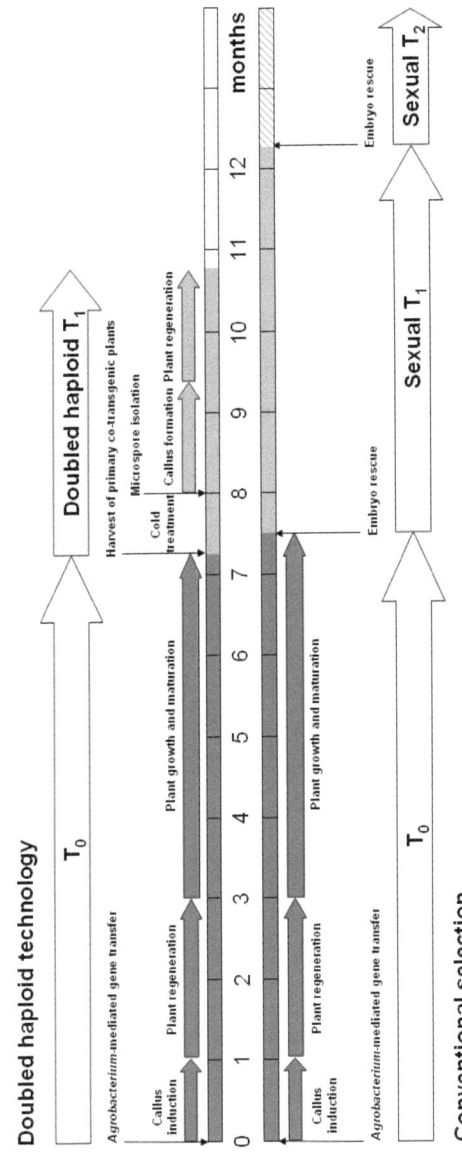

Diagram 4-1
Unequal efforts are required for the production and identification of selectable marker-free, GOI homozygous transgenic lines employing haploid technology as compared to conventional segregation in the most efficient method "two plasmids in one *Agrobacterium* clone" (variants 9 and 10)

DISCUSSION

Altogether the 'two plasmids in one *Agrobacterium* clone' method generated 39 independent primary transgenic lines, of which fourteen involved co-transformation (Table 3-2). Of these, doubled haploid progeny lacking the selectable marker but containing the GOI were produced from seven plants. The major reason for the high proportion of SM-free transgenic plants appears to be the predominantly uncoupled integration of the two different types of T-DNA. Even if the T-DNA copy number was high, the same T-DNA type was mostly integrated in linkage groups and behaved as one single locus often segregating independently from the other type of T-DNA.

The advantage of applying doubled haploid technology is that after identification of the selectable marker-free transgenic regenerants, there is no need to further test the plants with regard to their homozygosity, because the regenerants (through spontaneous or induced genome doubling) are instantly homozygous for the transgene. This means a significant reduction in the required greenhouse area in comparison to the conventional method using sufficiently high numbers of sexually produced T_2 lines to select SM-free true-breeding T_1 plants.

In comparison, in sexual T_1 populations with Mendelian segregation of unlinked T-DNAs, most of the plants tested positive for the GOI and lacking the selectable marker are hemizygous for the transgene, which requires a follow-up selection of homozygous lines that are ultimately needed for breeding purposes. To select the homozygous ones among the others the T_2 generations must be tested for the presence or absence of segregation event in each line. Not only remarkably more greenhouse space is needed in this case, but an enormous number of individuals need to be analysed to obtain the comparatively rare events of selectable marker free lines which are homozygous for the gene-of-interest.

It must be emphasized as well, that the established method must be robust and reproducible, with high proportion of unlinked T-DNA integration. However, agrobacteria rather prefer the integration of transgenes in linkage groups. An alternative

DISCUSSION

to span this problem would be the introduction of only the gene-of-interest without use of any selectable marker, as reported by Holme et al. (2006). They obtained 0.8 stable transgenic barley cv. 'Golden Promise' plants per 100 isolated ovules inoculated with agrobacteria without selective conditions.

Another aspect showing the relevance of this system, is that the European Deliberate Release Directive (2001/18/EC) requires the phasing out antibiotic resistance markers in GMOs, thus only transgenic lines free of the selectable marker are allowed to be used in field-test trials. Haploid technology not only provides homozygous transgenic lines, but is able to enhance the process of generating selectable marker-free progeny in the same time.

DISCUSSION

4.2. Integration of recombinant DNA in the barley genome

The reason why *Agrobacterium*-mediated gene transfer was chosen for the introduction of T-DNAs in the plant genome, was that the method results in transgenic plants of higher quality with regard to copy number and integrity of the transgene sequence as compared to direct DNA-transfer (Hensel et al. 2008; Kumlehn et al. 2006; Lange et al. 2006; Stahl et al. 2002; Travella et al. 2005).

The literature suggests that typically, the process of Agrobacterium-mediated DNA-transfer applied to immature barley embryos produces between one and three T-DNA inserts per event, with only around 10% of events involving four or more insertions (Hensel et al. 2008; Lange et al. 2006; Travella et al. 2005). Only small proportion (9-15%) of the transgenic lines are reported to contain 4 or more insertions (Hensel et al. 2008; Lange et al. 2006), although in 50% the T-DNA copies were integrated at the same locus (Stahl et al. 2002). About one half of all multiple inserts involve tandem or head-to-head insertions at a single site (Stahl et al. 2002).

Across the analysed set of 41 regenerants in the present experiments, the *gus* transgene copy number was from one to two in ca. 66% of the primary co-transgenic individuals, although the average copy number of the *hpt::gfp* sequence was rather higher (Diagram 3-4).

The T-DNA copy number itself does not give any information about the number of integration loci, e.g. a high number of transgene copies is not necessarily associated with many integration loci. Pursuing the pattern of transgene segregation in the DH and sexual T_1 populations is more conclusive (Table 3-5).

This can indicate if there are loci where more than one transgene copy was integrated in the genomic DNA of the plant cell by the gene transfer apparatus of *Agrobacterium*. In fact, multiple T-DNAs were often integrated linked to each other in the plant genome and behaved like a single locus (Hensel et al. 2008). In the present study, such linkage groups were frequently found among the variants (Figure 3-10). The failure to recover

DISCUSSION

selectable marker-free *gus* positive segregants from the Twin vectors generated co-transformants suggests that these vectors favour the integration of the two T-DNAs at a single site, possibly owing to frequent misinterpretation of the two adjacent T-DNAs as a single one by the gene transfer machinery of *Agrobacterium*. The highest number of independently behaving integration sites was found in variant 13, LBA4404/pTwin T, where eight transgenic fragments were integrated at five independent loci (Table 3-5). The chance for the segregation of the GOI might be very low if there were many unlinked SMs integrated at separate loci. The phenomenon that segregation of both of the selectable marker and the gene-of-interest occurred only in four lines among all variants, indicates that linked integration of transgenes is to be regarded as the typical phenomenon rather than uncoupled integration.

4.3. Further characteristics of the immature barley genetic transformation and DH production

The number of transgenic regenerants varied strongly between the three repeats of the experiments. The type of construct integrated in the plant genome might highly influence the transformation efficiency. Fang et al. (2002) reported that strong *gfp* expression reduced regeneration. Murray et al. (2004) compared *Agrobacterium*-mediated transformation of four barley genotypes and found that cells expressing *gfp* produced fewer regenerants than those expressing *gus*.

De Block and Debrouwer (1991) suggested that the identity of *Agrobacterium* strain used in the genetic transformation experiment might influence the pattern of T-DNA insertions in the plant genome; specifically, nopaline-derived strains such as AGL-1 tend to favour linked insertions, while octopine-derived ones, such as LBA4404 tend to favour unlinked ones. Matthews et al. (2001) failed to conduct genetic transformation of immature barley embryos using a Twin construct in LBA4404. The results presented in this thesis suggest, that the pattern of T-DNA integration seem to depend on both the

DISCUSSION

applied construct and the *Agrobacterium* strain. When using strain mixtures and the two plasmids in one *Agrobacterium* method, segregation of the gene-of-interest was a relatively frequent event. In contrast, the Twin variants tended to favour high *hpt::gfp* transgene copy numbers, and no event of GOI segregation was observed. The two T-DNAs, integrated in the Twin binary vectors in the present experiments were separated with only a short 500 bp sequence. As already mentioned by Matthews et al. (2001), a comparatively lower segregation frequency might be due to the proximity of the two T-DNAs on the transformation plasmid, so the selectable marker gene might be often integrated in a linked manner to the model gene-of-interest in the Twin variants. It is possible that increasing the length of the spacer sequence separating the two T-DNAs may have improved the chances of obtaining independent insertions.

In the present study, some 2% of the inoculated embryos produced more than one transgenic regenerant. Plants regenerated from the same callus may well not be clonal. According to a report, lines may arise from independent transformation events in a single co-cultured callus in rice (Sallaud et al. 2003). DNA gel blot-based profiling of the present materials identified both copy number and fragment size variation between those 'sister' regenerants (Table 3-1). The analysis of 101 regenerants derived from 20 of those embryos revealed that 6 of them had given rise to more than one genetically independent transgenic line, i.e. a total of 28 out of the 101 transgenic regenerants proved to be independent. As a result, a sensible routine practice would be to retain only one regenerant per explant.

Somaclonal variation occurs due to transformation and tissue culture processes (Bregitzer et al. 1998; Choi et al. 2001; Lemaux et al. 1999). Barley genetic transformation based on biolistics frequently produces tetraploid regenerants, which look abnormal and are partially sterile (Choi et al. 2002; Choi et al. 2003; Manoharan and Dahleen 2002).

DISCUSSION

At least 10% of the primary co-transgenic regenerants obtained in the present study proved tetraploid (Table 3-4). They included some which generated *gus* positive, *hpt* sensitive progeny. Microspore isolation from the spikes of such tetraploid barley plants was possible, and successful cultures producing green regenerants were obtained. Embryogenic pollen culture derived progeny from the tetraploids were diploid, but not necesseraly homozygous. Tetraploid (doubled dihaploids) plants were obtained as well, following spontaneous genome doubling. These T_1 regenerants could be tested in a successive round of transformation in order to assess their amenability for *Agrobacterium*-mediated gene transfer.

In case of one plant stemming from the two plamids/two strains method, 70% LBA4404/p*hpt::gfp* + 30% AGL-1/p*gus* (variant 8), an unstable transformation event occurred. Five copies of the selectable marker were detected in the primary co-transgenic plants, whereas none of them were found in the DH progeny. The integrated T-DNAs were probably integrated in a linkage group in one single locus. This indicates that the primary transgenic plant was chimeric with regard to the *hpt::gfp* insertion locus at which all of these five copies were likely to be linked (Marcotrigiano 1986), and that *hpt::gfp*-positive tissue of this chimera has not been involved in the formation of the spikes used to generate DH-lines. Another possibility might be that the integration site was a telomeric region of one of the chromosomes, where the loss of genetic information occures with a higher probability (Choi et al. 2002).

DISCUSSION

4.4. Identification of factors influencing the DH production efficiency

There were various conditions which have huge influence on the quality of callus growth, regeneration, the obtained number of transgenic plants and hereby the transformation frequency. Obviously, transgenic regenerants show phenotypic variability. Environmental conditions, quality of the donor plants and seasonal variability, transferred T-DNA sequence were factors which have a major effect on the outcome of plant genetic transformation (Hensel et al. 2008).

In the present study, the highest number of green regenerants was obtained in spring time, i.e. 1.6 doubled haploid lines were obtained per spike. However, it must be mentioned that the proportion of albinos, which are useless for any application, display a considerable sum (43.1%, data not shown). Throughout a whole year, between 9.3-17.6 doubled haploid T_1 plants are to be expected on average per 10.7 harvested spikes. The proportion of albino regenerants in the embryogenic pollen cultures was the lowest in summer (23.8%), highest in fall (49.1%). Their number might be lowered by optimisation of the culture conditions (Coronado et al. 2005). Genotype has a major impact on albino formation e.g embryogenic pollen cultures of the barley cultivar 'Igri' produce predominantly green regenerants.

On the other hand, quality of the donor material is the most influential on the outcome of the genetic transformation experiment and the number of regenerants of the embryogenic pollen cultures, e.g. in case of one of the most successful variants, LBA4404/p*hpt::gfp* + p*gus*), all the primary co-transgenic plants producing GOI positive antibiotic sensitive progeny are obtained from one single experiment. Matthews et al. (2001) reported about similar observations, in their case eight of their selectable marker free transgenic lines resulted from two experiments and already suggested that various factors, e.g. bacterial concentration and tissue culture conditions, might influence transformation and co-transformation efficiency.

5. ACKNOWLEDGMENTS

This work was done in the Leibniz Institute of Plant Genetics and Crop Plant Research (IPK) in Gatersleben.

I am grateful to my supervisor, Jochen Kumlehn, and to my mentor, Helmut Bäumlein, for the support during the time of my PhD studies.

I acknowledge the following people, who helped in different ways: Götz Hensel and David Köszegi for general advice about molecular work, Armin Meister for helping in statistical evaluation of the data, and Isolde Saalbach for valuable discussions.

I am especially thankful to Cornelia Marthe, Ingrid Otto and Sandra Wolf for excellent technical assistance.

I acknowledge to Jochen Kumlehn, Götz Hensel and David Köszegi for critical reading of the manuscript and for useful comments.

6. REFERENCES

Ahlandsberg S, Sathish P, Sun C, Jansson C (1999) Green fluorescent protein as a reporter system in the transformation of barley cultivar. Physiol Plant 107:194–200

Baker B, Schell J, Lorz H, Fedoroff N (1986) Transposition of the maize controlling element "Activator" in tobacco. Proc Natl Acad Sci U S A 83 (13):4844-4848

Bayley CC, Morgan M, Dale EC, Ow DW (1992) Exchange of gene activity in transgenic plants catalyzed by the Cre-lox site-specific recombination system. Plant Mol Biol 18 (2):353-361

Belzile F, Lassner MW, Tong Y, Khush R, Yoder JI (1989) Sexual transmission of transposed activator elements in transgenic tomatoes. Genetics 123 (1):181-189

Bregitzer P, Halbert SE, Lemaux PG (1998) Somaclonal variation in the progeny of transgenic barley. Theor Appl Genet 96:421-425

Carlson AR, Letarte J, Chen J, Kasha KJ (2001) Visual screening of microspore-derived transgenic barley (Hordeum vulgare L.) with green fluorescent protein. Plant Cell Rep 20:331–337

Chiu W, Niwa Y, Zeng W, Hirano T, Kobayashi H, Sheen J (1996) Engineered GFP as a vital reporter in plants. Curr Biol 6 (3):325-330

Cho MJ, Choi HW, Buchanan BB, Lemaux PG (1999) Inheritance of tissue-specific expression of barley hordein promoter-uidA fusions in transgenic barley plants. Theor Appl Genet 98:1253–1262

Cho MJ, Jiang W, Lemaux PG (1998) Transformation of recalcitrant barley cultivars through improvement of regenerability and decreased albinism. Plant Sci 138:229-244

Choi HW, Lemaux PG, Cho MJ (2000) Increased Chromosomal Variation in Transgenic versus Nontransgenic Barley (Hordeum vulgare L.) Plants. Crop Science 40:524-533

Choi HW, Lemaux PG, Cho MJ (2001) Selection and osmotic treatment exacerbate cytological aberrations in transformed barley (Hordeum vulgare). J Plant Physiol 158:935-943

Choi HW, Lemaux PG, Cho MJ (2002) Use of fluorescence in situ hybridization for gross mapping of transgenes and screening for homozygous plants in transgenic barley (Hordeum vulgare L.). Theor Appl Genet 106 (1):92-100

Choi HW, Lemaux PG, Cho MJ (2003) Long-term stability of transgene expression driven by barley endosperm-specific hordein promoters in transgenic barley. Plant Cell Rep 21:1108–1120

Cistué L, Ziauddin A, Simion E, Kasha KJ (1995) Effects of culture conditions on isolated microspore response of barley cultivar Igri. Plant Cell Tissue Organ Cult 42:163–169

Clapham D (1973) Haploid Hordeum plants from anthers in vitro. Z Pflanzenzüchtung 69:142-155

Coronado MJ, Hensel G, Broeders S, Otto I, Kumlehn J (2005) Immature pollen derived doubled haploid formation in barley cv. "Golden Promise" as a tool for transgene recombination. Acta Physiologiae Plantarum 27:591-599

Cotsaftis O, Sallaud J, Breitler JC, Meynard D, Greco R, Pareira A, Guiderdoni E (2002) Transposon-mediated generation of T-DNA and marker free rice plants expressing a Bt endotoxin gene. Mol Breed 10:165-180

Dale EC, Ow DW (1991) Gene transfer with subsequent removal of the selection gene from the host genome. Proc Natl Acad Sci U S A 88 (23):10558-10562

Daley M, Knauf VC, Summerfelt KR, Turner JC (1998) Co-transformation with one Agrobacterium tumefaciens strain containing two binary plasmids as a method for producing marker-free transgenic plants. Plant Cell Rep 17:489–496

De Block M, Debrouwer D (1991) Two T-DNAs cotransformed into Brassica napus by a double Agrobacterium tumefaciens infection are mainly integrated at the same locus. Theor Appl Genet 82:257–263

De Framond AJ, Back EW, Chilton WS, Kayes L, Chilton M (1986) Two unlinked T-DNAs can transform the same tobacco plant cell and segregate in the F1 generation. Mol Gen Genet 202:125–131

Depicker A, Herman L, Jacobs A, Schell J, Van Montagu M (1985) Frequencies of simultaneous transformation with different T-DNAs and their relevance to the Agrobacterium/ plant interaction. Mol Gen Genet 201:477–484

Dunwell JM (1985) Anther and ovary culture. Cereal tissue and cell culture. Dordrecht,

Ebinuma H, Sugita K, Matsunaga E, Yamakado M (1997) Selection of marker-free transgenic plants using the isopentenyl transferase gene. Proc Natl Acad Sci U S A 94 (6):2117-2121

Endo S, Sugita K, Sakai M, Tanaka H, Ebinuma H (2002) Single-step transformation for generating marker-free transgenic rice using the ipt-type MAT vector system. Plant J 30 (1):115-122

Erikson O, Hertzberg M, Nasholm T (2004) A conditional marker gene allowing both positive and negative selection in plants. Nat Biotechnol 22 (4):455-458

Fang YD, Akula C, Altpeter F (2002) Agrobacterium-mediated barley (Hordeum vulgare L.) transformation using green fluorescent protein as a visual marker and sequence analysis of the T-DNA:genomic DNA junctions. J Plant Physiol 159

Forster BP, Pakniyat H, Macaulay M, Matheson W, Phillips MS, Thomas WTB, Powell W (1994) Variation in the leaf sodium content of Hordeum vulgare (barley) cultivar Maythorpe and its derived mutant cv. "Golden Promise". Heredity 73:249–253

Funatsuki H, Kuroda H, Kihara M, Lazzeri PA, Miiller E, Lörz H, Kishinami I (1995) Fertile transgenic barley generated by direct DNA transfer to protoplasts Theor Appl Genet 91:707-712

Gaponenko AK, Petrova TF, Iskakov AR, Sozinov AA (1988) Cytogenetics of in vitro cultured somatic cells and regenerated plants of barley (Hordeum vulgare L.). Theor Appl Genet 75:905-911

Garfinkel DJ, Nester EW (1980) Agrobacterium tumefaciens mutants affected in crown gall tumorigenesis and octopine catabolism. J Bacteriol 144:732-743

Gleave AP, Mitra DS, Mudge SR, Morris BA (1999) Selectable marker-free transgenic plants without sexual crossing: transient expression of cre recombinase and use of a conditional lethal dominant gene. Plant Mol Biol 40 (2):223-235

Goedeke S, Hensel G, Kapusi E, Gahrtz M, Kumlehn J (2007) Transgenic Barley in Fundamental Research and Biotechnology. Transgenic Plant J 1:104-117

Goldsbrough AP, Lastrella CN, Yoder JI (1993) Transposition mediated re- positioning and subsequent elimination of marker genes from transgenic tomato. Biotechnology 11:1286-1292

Gorbunova V, Levy AA (2000) Analysis of extrachromosomal Ac/Ds transposable elements. Genetics 155 (1):349-359

Gustafson VD, Baenziger PS, Wright MS, Stroup WW, Yen Y (1995) Isolated wheat microspore culture. Plant Cell Tiss Org Cult 42:207-213

Hagberg A, Hagberg G (1980) High frequency of spontaneous haploids in the progeny of an induced mutation barley. Hereditas 93:341-343

Hagio T, Hirabayshi T, Machii H, Tomotsune H (1995) Production of fertile transgenic barley (Hordeum vulgare L.) plants using the hygromycin-resistance gene. Plant Cell Reports 14:329-334

Harwood WA, Ross SM, Bulley SM, Travella S, Busch B, Harden J, Snape JW (2002) Use of the firefly luciferase gene in a barley (Hordeum vulgare) transformation system. Plant Cell Rep 21:320-326

Hellens R, Mullineaux P, Klee H (2000) Technical Focus:a guide to Agrobacterium binary Ti vectors. Trends Plant Sci 5 (10):446-451

Hensel G, Kumlehn J (2004) Genetic transformation of barley (Hordeum vulgare L.) by co-culture of immature embryos with Agrobacteria. In: Curtis IS (ed) Transgenic crops of the world-essential protocols. Kluwer, Dordrecht, pp 35-45

Hensel G, Kumlehn J (2009) Genetic transformation technology in the Triticeae. Breeding Science 59:553-560

Hensel G, Valkov V, Middlefell-Williams J, Kumlehn J (2008) Efficient generation of transgenic barley: the way forward to modulate plant-microbe interactions. J Plant Physiol 165 (1):71-82

Hoa TT, Bong BB, Huq E, Hodges TK (2002) Cre/ lox site-specific recombination controls the excision of a transgene from the rice genome. Theor Appl Genet 104 (4):518-525

Hoekstra AP, van Zijderveld MM, Louwerse JD, Heidekamp F, van der Mark F (1992) Anther and microspore culture of Hordeum vulgare L. cv. Igri. Plant Sci 86:89-96

Hohn B, Levy AA, Puchta H (2001) Elimination of selection markers from transgenic plants. Curr Opin Biotech 12:139-143

Holm PB, Olsen O, Schnorf M, Brinch-Pedersen H, Knudsen S (2000) Transformation of barley by microinjection into isolated zygote protoplasts. Transgenic Research 9:21-32

Holme IB, Brinch-Pedersen H, Lange M, Holm PB (2006) Transformation of barley (Hordeum vulgare L.) by Agrobacterium tumefaciens infection of in vitro cultured ovules. Plant Cell Rep 25 (12):1325-1335

Hu TC, Kasha KJ (1997) Improvement of isolated microspore culture of wheat (Triticum aestivum L.) through ovary co-culture. Plant Cell Rep 16:520-525

Hunter CP (1987) Plant generation method

Indrianto A, Heberle-Bors E, Touraev A (1999) Assessment of various stresses and carbohydrates for their effect on the induction of embryogenesis in isolated wheat microspore culture. Plant Sci 143:71-79

Jähne A, Becker D, Brettschneider R, Lörz H (1994) Regeneration of transgenic, microspore-derived, fertile barley. Theor Appl Genet 89:525-533

Jefferson RA (1987) Assaying chimeric genes in plants by the GUS fusion system. Plant Mol Biol Rep 5:387-405

Jia H, Pang Y, Chen X, Fang R (2006) Removal of the selectable marker gene from transgenic tobacco plants by expression of cre recombinase from a tobacco mosaic virus vector through agroinfection. Transgenic Res 15:375-384

Kamal-Eldin K, Appelvist LA (1996) The chemistry and antioxidant properties of tocopherols and tocotrrienols. Lipids 31 (7):671-701

Kao NK (1993) Viability, cell division and microcallus formation of barley microspores in culture. Plant Cell Reports 12:366-369

Kasha KJ, Reinbergs E (1982) Recent developments in the production and utilization of haploids in barley. Proc 4th Int Barley Genet Symp. Edinburgh

Kasha KJ, Simion E, Oro R, Yao QA, Hu TC, Carlson AR (2001) An improved in vitro technique for isolated microspore culture of barley. Euphytica 120:379-385

Kasha KJ, Ziauddin A, Cho UH **XIX Stadler Genetics Symp, Missouri** In, 1989. pp 13-236

Kerbach S, Lorz H, Becker D (2005) Site-specific recombination in Zea mays. Theor Appl Genet 111 (8):1608-1616

Kilby NJ, Davies GJ, Snaith MR (1995) FLP recombinase in transgenic plants: constitutive activity in stably transformed tobacco and generation of marked cell clones in Arabidopsis. Plant J 8 (5):637-652

Kittiwongwattana C, Lutz K, Clark M, Maliga P (2007) Plastid marker gene excision by the phiC31 phage site-specific recombinase. Plant Mol Biol 64 (1-2):137-143

Köhler F, Wenzel G (1985) Regeneration of isolated barley microspores in conditioned media and trials to characterize the responsible factor. J Plant Physiol 121:181-191

Komari T, Hiei Y, Saito Y, Murai N, Kumashiro T (1996) Vectors carrying two separate T-DNAs for co-transformation of higher plants mediated by Agrobacterium tumefaciens and segregation of transformants free from selection markers. Plant J 10 (1):165-174

Kopertekh L, Jüttner G, Schliemann J (2004) Site-specific recombination induced in transgenic plants by PVX virus vector expressing bacteriophage P1 recombinase. Plant Sci 166:485-492

Koprek T, McElroy D, Louwerse J, Williams-Carrier R, Lemaux PG (1999) Negative selection systems for transgenic barley (Hordeum vulgare L.): comparison of bacterial codA- and cytochrome P450 gene-mediated selection. Plant J 19 (6):719-726

Kuhlmann V, Foroughi-Wehr B (1989) Production of doubled haploid lines in frequencies sufficient for barley breeding program. Plant Cell Reports 8:78-81

Kumlehn J, Lörz H (1999) Monitoring sporophytic development of individual microspores of barley (Hordeum vulgare L.). Anther and Pollen: From Biology to Biotechnology. Springer, Berlin Heidelberg New York

Kumlehn J, Serazetdinova L, Hensel G, Becker D, Loerz H (2006) Genetic transformation of barley (Hordeum vulgare L.) via infection of androgenetic pollen cultures with Agrobacterium tumefaciens. Plant Biotechnol J 4 (2):251-261

Kunze R (1996) The maize transposable element Activator (Ac), vol 24. In Current Topics in Microbiology and Immunology, Transposable Elements. Springer-Verlag,

Lange M, Vincze E, Moller MG, Holm PB (2006) Molecular analysis of transgene and vector backbone integration into the barley genome following Agrobacterium-mediated transformation. Plant Cell Rep 25 (8):815-820

Lassner MW, Palys JM, Yoder JI (1989) Genetic transactivation and Dissociation elements in transgenic tomato plants. Mol Gen Genet 218:25-32

Lazo GR, Stein PA, Ludwig RA (1991) A DNA transformation-competent Arabidopsis genomic library in Agrobacterium. Biotechnology (N Y) 9 (10):963-967

Lazzeri PA, Brettschneider R, Liihrs R, Lörz H (1991) Stable transformation of barley via PEG-induced direct DNA uptake into protoplasts. Theor Appl Genet 81:437-444

Lemaux PG, Cho MJ, Zhang S, Bregitzer P (1999) Transgenic cereals: Hordeum vulgare L. (barley). In. Kluwer, Dordrecht,

Li H, Devaux P (2001) Enhancement of microspore culture efficiency of recalcitrant barley genotypes. Plant Cell Report 20:475-481

Li H, Devaux P (2003) High frequency regeneration of barley doubled haploid plants from isolated microspore culture. Plant Sci 164:379-386

Luckett DJ (1989) Colchicine mutagenesis is associated with substantial heritable variation in cotton. Euphytica 42:177–182

Lyznik LA, Rao KV, Hodges TK (1996) FLP-mediated recombination of FRT sites in the maize genome. Nucleic Acids Res 24 (19):3784-3789

Manoharan M, Dahleen DS (2002) Genetic transformation of the commercial barley Hordeum vulgare L.) cultivar Conlon by particle bombardment of callus. Plant Cell Rep 21:76–80

Marcotrigiano M (1986) Origin of adventitious shoots regenerated from cultured tobacco leaf tissue. Amer J Bot 73:1541-1547

Masterson RV, Furtek DB, Grevelding C, Schell J (1989) A maize Ds transposable element containing a dihydrofolate reductase gene transposes in Nicotiana tabacum and Arabidopsis thaliana. Mol Gen Genet 219:461-466

Matthews PR, Wang MB, Waterhouse PM, Thornton S, Fieg SJ, Gubler F, Jacobsen JV (2001) Marker gene elimination from transgenic barley, using co-transformation with adjacent "twin T-DNAs" on a standard Agrobacterium transformation vector. Mol Breed 7:195-202

McCormac AC, Wu H, Bao M, Wang Y, Xue R, Elliott MC, Chen DF (1998) The use of visual marker genes as cell-specific reporters of Agrobacterium-mediated T-DNA delivery to wheat (Triticum aestivum L.) and barley (Hordeum vulgare L.). Euphytica 99:17–25

McElroy D, Zhang W, Cao J, Wu R (1990) Isolation of an efficient actin promoter for use in rice transformation. Plant Cell 2 (2):163-171

McKnight TD, Lillis MT, Simpson RB (1987) Segregation of genes transferred to one plant cell from two separate Agrobacterium strains. Plant Mol Biol 8:439–445

Mejza SJ, Morgant V, DiBona D, Wong JR (1993) Plant regeneration from isolated micro spores of Triticum aestivum. Plant Cell Rep 12:149-153

Miki B, McHugh S (2004) Selectable marker genes in transgenic plants: applications, alternatives and biosafety. J Biotechnol 107 (3):193-232

Mordhorst AP, Lörz H (1993) Embryogenesis and development of isolated barley (Hordeum vulgare L.) micro spores are influenced by the amount and composition of nitrogen sources in culture media. J Plant Physiol 142:485-492

Murray F, Brettell R, Matthews P, Bishop D, Jacobsen J (2004) Comparison of Agrobacterium-mediated transformation of four barley cultivars using the GFP and GUS reporter genes. Plant Cell Rep 22 (6):397-402

Nobre J, Davey MR, Lazzeri PA, Cannell ME (2000) Transformation of barley scutellum protoplasts: regeneration of fertile transgenic plants. Plant Cell Reports 19:1000–1005

Nuutila AM, Ritala A, Skadsen RW, Mannonen L, Kauppinen V (1999) Expression of fungal thermotolerant endo 1,4-β-glucanase in transgenic barley seeds during germination. Plant Mol Biol 41:777-783

Odell J, Caimi P, Sauer B, Russell S (1990) Site-directed recombination in the genome of transgenic tobacco. Mol Gen Genet 223 (3):369-378

Odell JT, Nagy F, Chua NH (1985) Identification of DNA sequences required for activity of the cauliflower mosaic virus 35S promoter. Nature 313 (6005):810-812

Olsen FL (1991) Isolation and cultivation of embryogenic microspores from barley (Hordeum vulgare L.). Hereditas 115:255-266

Onouchi H, Nishihama R, Kudo M, Machida Y, Machida C (1995) Visualization of site-specific recombination catalyzed by a recombinase from Zygosaccharomyces rouxii in Arabidopsis thaliana. Mol Gen Genet 247 (6):653-660

Ow DW (2007) GM maize from site-specific recombination technology, what next? Curr Opin Biotechnol 18 (2):115-120

Palotta MA, Graham RD, Langridge P, Sparrow DHB, Barker SJ (2000) RFLP mapping of manganese efficiency in barley. Theor Appl Genet 101:1100-1108

Patel M, Johnson JS, Brettell RIS, Jacobsen J, Xue GP (2000) Transgenic barley expressing a fungal xylanase gene in the endosperm of the developing grains. Mol Breed 6:113-123

Pickering RA, Devaux P (1992) Haploid production: approaches and use in plant breeding. In. Oxford: Alden Press Ltd,

Puchta H (2000) Removing selectable marker genes: taking the shortcut. Trends Plant Sci 5 (7):273-274

Reed J, Privalle L, Powell ML, Meghji M, Dawson J, Dunder E, Suttie J, Wenck A, Launis K, Kramer C, Chang YF, Hansen G, Wright M (2001) Phosphomannose isomerase: an efficient selectable marker for plant transformation. In Vitro Cell Dev Biol Plant 37:127-132

Ritala A, Aspegren K, Kurten U, Salmenkallio-Marttila M, Mannonen L, Hannus R, Kauppinen V, Teeri TH, Enari TM (1994) Fertile transgenic barley by particle bombardment of immature embryos. Plant Mol Biol 24:317-325

Ritala A, Mannonen L, Oksman-Caldentey KM (2001) Factors affecting the regeneration capacity of isolated barley microspores (Hordeum vulgare L.). Plant Cell Report 20:403-407

Russell SH, Hoopes JL, Odell JT (1992) Directed excision of a transgene from the plant genome. Mol Gen Genet 234 (1):49-59

Sallaud C, Meynard D, van Boxtel J, Gay C, Bes M, Brizard JP, Larmande P, Ortega D, Raynal M, Portefaix M, Ouwerkerk PB, Rueb S, Delseny M, Guiderdoni E (2003) Highly efficient production and characterization of T-DNA plants for rice (Oryza sativa L.) functional genomics. Theor Appl Genet 106 (8):1396-1408

Salmenkallio-Marttila M, Aspegren K, Åkerman S, Kurtén U, Mannonen L, Ritala A, Teeri TH, Kauppinen V (1995a) Transgenic barley (Hordeum vulgare L.) by electroporation of protoplasts. Plant Cell Rep 15:301–304

Salmenkallio-Marttila M, Kurtén U, Kauppinen V (1995b) Culture conditions for efficient induction of green plants from isolated microspores of barley. Plant Cell Tissue Organ Cult 43:79–81

Salvo-Garrido H, Travella S, Bilham LJ, Harwood WA, Snape JW (2004) The distribution of transgene insertion sites in barley determined by physical and genetic mapping. Genetics 167 (3):1371-1379

Sambrook J, Fritsc EF, Maniatis T (1989) Molecular cloning - a laboratory manual. New York: Cold Spring Laboratory Press,

Schledzewski K, Mendel R (1994) Quantitative transient gene expression: comparison of the promoters for maize polyubiquitin 1 rice actin 1, maize derived Emu and CaMV 35S in cells of barley, maize and tobacco. Transgenic Research 3:249–255

Scott P, Lyne RL (1994) The effect of different carbohydrate sources upon the initiation of embryogenesis from barley microspores. Plant Cell Tiss Org Cult 36:129-133

Shrawat AK, Lörz H (2006) Agrobacterium-mediated transformation of cereals: a promising approach crossing barriers. Plant Biotechnology Journal 4:575-603

Silhavy TJ, Berman ML, Enquist LW (1984) Experiments with gene fusions. Cold Spring Harbor Laboratory Press, Cold Spring Harbor, N.Y.

Srivastava V, Ow DW (2003) Rare instances of Cre-mediated deletion product maintained in transgenic wheat. Plant Mol Biol 52 (3):661-668

Stahl R, Horvath H, Van Fleet J, Voetz M, von Wettstein D, Wolf N (2002) T-DNA integration into the barley genome from single and double cassette vectors. Proc Natl Acad Sci U S A 99 (4):2146-2151

Sugita K, Kasahara T, Matsunaga E, Ebinuma H (2000) A transformation vector for the production of marker-free transgenic plants containing a single copy transgene at high frequency. Plant J 22 (5):461-469

Sunderland N, Roberts M, Evans LJ, Wildon DC (1978) Multicellular pollen formation in cultured barley anthers. J Exp Bot 30:1133–1144

Sunderland N, Xu ZH (1982) Shed pollen culture in Hordeum vulgare. J Exp Bot 33:1086

Takamura T, Miyajima I (1996) Colchicine induced tetraploids in yellow-flowered cyclamens and their characteristics. Scientia Horticulturae 65 (4):305–312

Thiebaut J, Kasha KJ (1978) Modification of the Colchicine Technique for Chromosome Doubling of Barley Haploids. Can J Genet Cytol 20:513-521

Thorpe HM, Smith MC (1998) In vitro site-specific integration of bacteriophage DNA catalyzed by a recombinase of the resolvase/invertase family. Proc Natl Acad Sci U S A 95 (10):5505-5510

Tingay S, McElroy D, Kalla R, Feig S, Wang M, Thornton S, Brettell R (1997) Agrobacterium tumefaciens-mediated barley transformation. Plant J 11:1369-1376

Touraev A, Indrianto I, Wratschko I, Vicente O, Heberle-Bors E (1996) Efficient microspore embryogenesis in wheat (Triticum aestivum L.) induced by starvation at high temperature. Sex Plant Reprod 9:209-215

Touraev A, Vicente O, Heberle-Bors E (1997) Initiation of microspore embryogenesis by stress. Trends Plant Sci 2:297-302

Travella S, Ross SM, Harden J, Everett C, Snape JW, Harwood WA (2005) A comparison of transgenic barley lines produced by particle bombardment and Agrobacterium-mediated techniques. Plant Cell Rep 23 (12):780-789

Trifonova A, Madsen S, Olesen A (2001) Agrobacterium-mediated transgene delivery and integration into barley under a range of in vitro culture conditions. Plant Sci 162:871-880

Vancanneyt G, Schmidt R, O'Connor-Sanchez A, Willmitzer L, Rocha-Sosa M (1990) Construction of an intron-containing marker gene: splicing of the intron in transgenic plants and its use in monitoring early events in Agrobacterium-mediated plant transformation. Mol Gen Genet 220 (2):245-250

Wan Y, Lemaux PG (1994) Generation of Large Numbers of Independently Transformed Fertile Barley Plants. Plant Physiol 104 (1):37-48

Wang L, Xue Q, Newman RK, Newman CW (1993) Enrichment of tocopherol, tocotrienol and oil in barley by milling and pearling. Cereal Chemistry 70 (5):499-501

Wang MB, Abbott DC, Upadhyaya NM, Jacobsen JV, Waterhouse PM (2001) Agrobacterium tumefaciens-mediated transformation of an elite Australian barley cultivar with virus resistance and reporter genes. Aust J Plant Physiol 28:149-156

Wang MB, Waterhouse PM (1997) A rapid and simple method of assaying plants transformed with hygromycin or PPT resistance genes. Plant Mol Biol Rep 15:209-215

Wu H, McCormac AC, Elliott MC, Chen DF (1998) Agrobacterium-mediated stable transformation of cell suspension cultures of barley (Hordeum vulgare). Plant Cell, Tissue and Organ Culture 54:161-171

Xue GP, Patel M, Johnson JS, Smyth DJ, Vickers CE (2003) Selectable marker-free transgenic barley producing a high level of cellulase (1,4-β-glucanase) in developing grains. Plant Cell Rep 21:1088-1094

Yoder JI, Goldsbrough AP (1994) Transformation systems for generating marker-free transgenic plants. Biotechnology 12:263-267

Zhang S, Cho MJ, Koprek T, Yun R, Bregitzer P, Lemaux PG (1999) Genetic transformation of commercial cultivars of oat (Avena sativa L.) and barley (Hordeum vulgare L.) using shoot meristematic cultures derived from germinated seedlings. Plant Cell Rep 18:959–966

Zhang W, Subbarao S, Addae P, Shen A, Armstrong C, Peschke V, Gilbertson L (2003) Cre/lox-mediated marker gene excision in transgenic maize (Zea mays L.) plants. Theor Appl Genet 107 (7):1157-1168

Zheng MY, Liu W, Weng Y, Polle E, Konzak CF (2001) Culture of freshly isolated wheat (Triticum aestivum L.) microspores treated with inducer chemicals. Plant Cell Rep 20:685-690

Ziauddin A, Marsolais A, Simion E, Kasha KJ (1992) Improved plant regeneration from wheat anther and barley microspore culture using phenylacetic acid (PAA). Plant Cell Reports 11:489-498

Ziauddin A, Simion E, Kasha KJ (1990) Improved plant regeneration from shed microspore culture in barley (Hordeum vulgave L.) cv. Igri. Plant Cell Reports 9:69-72

Zubko E, Scutt C, Meyer P (2000) Intrachromosomal recombination between attP regions as a tool to remove selectable marker genes from tobacco transgenes. Nat Biotechnol 18 (4):442-445

7. APPENDIX

Variant No.	No. of inoculated embryos	No. of transgenic regenerants (SM+)	Transgenic independent lines (SM+)	Escapes	No. of co-transgenic regenerants (SM+,GOI+)	Co-transgenic independent lines (SM+,GOI+)	Independent co-transgenic line(s) producing green DH progeny	Independent co-transgenic line(s) producing GOI positive marker free DH progeny	Sexual line(s) producing GOI positive marker free DH progeny
1	270	44	11	1	2	2	1	0	0
2	270	28	16	1	8	5	3	0	1
control 1,2	270	9	5	0	7	3	3	1	0
3	270	50	21	0	2	2	1	0	0
4	270	54	26	0	6	2	2	2	0
5	270	8	4	2	4	1	1	0	0
control 3,4,5	270	24	15	0	5	4	4	1	0
6	270	18	9	0	5	3	3	2	0
8	270	24	13	3	1	1	1	1	0
control 6,8	270	28	13	3	5	4	4	0	0
9	270	46	24	0	16	8	6	3	0
10	270	54	15	0	9	6	4	4	0
control 9,10	270	24	5	2	9	2	2	0	0
11	270	57	15	0	24	6	4	0	0
12	270	21	6	2	2	1	1	0	0
control 11,12	270	6	2	0	0	0	0	0	0
13	270	44	10	0	6	2	0	0	0
14	270	27	6	0	16	2	1	0	0
control 13,14	270	40	12	0	2	1	1	0	0
control sum	1620	131	52	5	28	14	14	2	0

i want morebooks!

Buy your books fast and straightforward online - at one of world's fastest growing online book stores! Environmentally sound due to Print-on-Demand technologies.

Buy your books online at
www.get-morebooks.com

Kaufen Sie Ihre Bücher schnell und unkompliziert online – auf einer der am schnellsten wachsenden Buchhandelsplattformen weltweit! Dank Print-On-Demand umwelt- und ressourcenschonend produziert.

Bücher schneller online kaufen
www.morebooks.de

VDM Verlagsservicegesellschaft mbH
Heinrich-Böcking-Str. 6-8
D - 66121 Saarbrücken

Telefon: +49 681 3720 174
Telefax: +49 681 3720 1749

info@vdm-vsg.de
www.vdm-vsg.de

Printed by Books on Demand GmbH, Norderstedt / Germany